Aliaksei Putau

Organocuprate Aggregation and Reactivity: Decoding the 'Black Box'

Aliaksei Putau

Organocuprate Aggregation and Reactivity: Decoding the 'Black Box'

Südwestdeutscher Verlag für Hochschulschriften

Impressum / Imprint

Bibliografische Information der Deutschen Nationalbibliothek: Die Deutsche Nationalbibliothek verzeichnet diese Publikation in der Deutschen Nationalbibliografie; detaillierte bibliografische Daten sind im Internet über http://dnb.d-nb.de abrufbar.
Alle in diesem Buch genannten Marken und Produktnamen unterliegen warenzeichen-, marken- oder patentrechtlichem Schutz bzw. sind Warenzeichen oder eingetragene Warenzeichen der jeweiligen Inhaber. Die Wiedergabe von Marken, Produktnamen, Gebrauchsnamen, Handelsnamen, Warenbezeichnungen u.s.w. in diesem Werk berechtigt auch ohne besondere Kennzeichnung nicht zu der Annahme, dass solche Namen im Sinne der Warenzeichen- und Markenschutzgesetzgebung als frei zu betrachten wären und daher von jedermann benutzt werden dürften.

Bibliographic information published by the Deutsche Nationalbibliothek: The Deutsche Nationalbibliothek lists this publication in the Deutsche Nationalbibliografie; detailed bibliographic data are available in the Internet at http://dnb.d-nb.de.
Any brand names and product names mentioned in this book are subject to trademark, brand or patent protection and are trademarks or registered trademarks of their respective holders. The use of brand names, product names, common names, trade names, product descriptions etc. even without a particular marking in this works is in no way to be construed to mean that such names may be regarded as unrestricted in respect of trademark and brand protection legislation and could thus be used by anyone.

Coverbild / Cover image: www.ingimage.com

Verlag / Publisher:
Südwestdeutscher Verlag für Hochschulschriften
ist ein Imprint der / is a trademark of
AV Akademikerverlag GmbH & Co. KG
Heinrich-Böcking-Str. 6-8, 66121 Saarbrücken, Deutschland / Germany
Email: info@svh-verlag.de

Herstellung: siehe letzte Seite /
Printed at: see last page
ISBN: 978-3-8381-3698-1

Zugl. / Approved by: München, LMU, Diss., 2012

Copyright © 2013 AV Akademikerverlag GmbH & Co. KG
Alle Rechte vorbehalten. / All rights reserved. Saarbrücken 2013

Acknowledgements

It is a pleasure to thank the many people who made this project possible.

My gratitude to my PhD supervisor, Prof. Dr. Konrad Koszinowski, is difficult to overstate. His sound advice, good company, and enthusiasm for trying new things ensured the tropical sunny climate in our group that is so necessary for fruitful (and, above all, enjoyable) research. Besides, his involvement in the group life outside the lab made all of us group members feel like team members, not just a loose bunch of colleagues.

I am also deeply indebted to Prof. Dr. Herbert Mayr, not only for his continuous generous support and help, but also for running such a cool, all-star research group, which provided a great many social occasions.

I thank SFB 749 for the financial support of this project, specifically Mrs. Birgit Carell, for her support and understanding of my worries and troubles of all kinds.

Furthermore, my sincere gratitude goes to Prof. Dr. Hendrik Zipse, Prof. Dr. Manfred Heuschmann, Prof. Dr. Konstantin Karaghiosoff and Prof. Dr. Paul Knochel for showing interest in this manuscript by accepting to be referees.

This work would have hardly been imaginable without my teammates, Katharina Böck, Christina Müller and Julia Fleckenstein, whose friendship, help and motivation made me feel at home.

Finally, last but not least, I would like to thank my wife and family for their endless love and support.

Table of Contents

1 **Abstract** 1

2 **Introduction** 3
   2.1 Overview 3
   2.2 Aggregation and Structure of Cyanocuprates 5
   2.3 Coupling Reactions 7
   2.4 Conjugate Addition Reactions 9
   2.5 Objectives 10

3 **Instrumentation and Methods** 11
   3.1 Electrospray Ionization Mass Spectrometry (ESI-MS) 11
      3.1.1 Theoretical Overview 11
      3.1.2 Potential and Limitations of ESI-MS 16
      3.1.2 Experimental Part 17
   3.2 Electrical Conductivity Measurements 23
   3.3 Theoretical Calculations on Tetraalkylcuprates(III) 25
   3.4 Analysis of Energy-Dependent Fragmentation Reactions 26

4 **Results and Discussion** 29
   4.1 Aggregation and Structure of Cyanocuprates 29
      4.1.1 Negative-Ion Mode ESI Mass Spectrometry 29
      4.1.2 Positive-Ion Mode ESI Mass Spectrometry 39
      4.1.3 Electrical Conductivity Measurements 42
      4.1.4 General Trends 45
      4.1.5 Equilibria Operative 46
      4.1.6 Effect of the Solvent 49
      4.1.7 Effect of the Organyl Substituent 51
      4.1.8 Effect of the Temperature 53
      4.1.9 Comparison of Analytical Methods 53
   4.2 Gas-Phase Reactivity of Cyanocuprates 55
      4.2.1 Gas-Phase Fragmentation Reactions 55
      4.2.2 Gas-Phase Hydrolysis Reactions 63

4.3 Cross-Coupling Reactions 69
    4.3.1 Reactions of Dialkylcuprates with Organyl Halides 69
    4.3.2 Association Equilibria of Lithium Tetraalkylcuprates 77
    4.3.3 Unimolecular Reactivity of Tetraalkylcuprates 82
4.4 Conjugate Addition Reactions 93
    4.4.1 Reactions of Diorganylcuprates with Acrylonitrile 93
    4.4.2 Reactions of Diorganylcuprates with Fumaronitrile 93
    4.4.3 Reactions of Diorganylcuprates with 1,1-Dicyanoethylene 96
    4.4.4 Reactions of Diorganylcuprates with Tricyanoethylene 96
    4.4.5 Reactions of Diorganylcuprates with Tetracyanoethylene 99
    4.4.5 Substrate Structure-Reactivity Relationships 102

5 Conclusions and Outlook 104

6 Appendix 108
6.1 Analytical Data 108
6.2 Synthesis 109
    6.2.1 General Considerations 109
    6.2.2 Synthesis of Organocuprate Reagents 109
    6.2.3 Synthesis of Cu^{13}CN 111
    6.2.4 Synthesis of Cyanoethylene Substrates 112
6.3 Determination of Background Water Concentration 115

7 References and Notes 116

1 Abstract

A broad range of organocopper intermediates in different aggregation states were characterized by electrospray ionization (ESI) mass spectrometry, which provided valuable information on these fluxional species. To complement the mass spectrometric data, electrical conductivity measurements and theoretical calculations were employed.

Tetrahydrofuran (THF) solutions of CuCN/(RLi)$_m$ stoichiometry (m = 0.5, 0.8, 1.0, and 2.0 and R = Me, Et, ^{n}Bu, ^{s}Bu, ^{t}Bu, Ph) were analyzed by ESI mass spectrometry, and organocuprate anions were detected for all cases. The composition of these species showed clear dependence on the amount of RLi used. Thus, while cyanide-free Li$_{n-1}$Cu$_n$R$_{2n}^-$ anions completely predominated for CuCN/(RLi)$_2$ solutions, cyanide-containing Li$_{n-1}$Cu$_n$R$_n$(CN)$_n^-$ complexes prevailed for CuCN/(RLi)$_m$ reagents with $m \leq 1$. Ligand mixing studies on LiCuMe$_2$·LiCN and LiCuR$_2$·LiCN systems (R = Et, ^{n}Bu, ^{s}Bu, ^{t}Bu, Ph) revealed fast exchange equilibria operating in solution.

When THF was substituted for the less polar diethyl ether (Et$_2$O), no major new species were observed. However, the proportion of higher nuclearity anions was consistently greater in the latter solvent than in the former. Further experiments with 2-methyltetrahydrofuran (MeTHF), cyclopentyl methyl ether (CPME) and methyl *tert*-butyl ether (MTBE) solutions confirmed the suggestion that higher aggregation states are favored by lower polarity solvents. Additional conductivity experiments indicated that contact ion pairs strongly predominate for solutions in Et$_2$O, whereas the more polar THF gives rise to larger amounts of solvent-separated ion pairs.

Following the detection of organocuprate ions, their gas- and condensed-phase reactions were investigated. Collision-induced dissociation (CID) experiments were used to study intrinsic reactivities in the gas phase. Higher aggregates were found to break apart into fragments of lower nuclearity, whereas monomeric species decomposed by β-H elimination when possible. In some CID spectra, the presence of hydroxyl-containing signals led to the conclusion that a reaction with background water inside the mass spectrometer was taking place. This bimolecular reaction was then studied in detail for many different systems. The results indicate that lithium centers seem to be a necessary (but not only) pre-requisite for hydrolysis.

1 Abstract

For example, no reaction was observed for monomeric CuMe$_2^-$ anions, whereas the reactions of LiCu$_2$Me$_4^-$ and Li$_2$Cu$_3$Me$_6^-$ were much faster.

Following the successful characterization of organocuprates, their synthetically useful coupling reactions with alkyl halides were probed. ESI mass spectrometric experiments, supported by electrical conductivity measurements, indicated that LiCuMe$_2$·LiCN reacts with a series of alkyl halides RX (R = Me, Et, ^{n}Pr, ^{n}Bu, PhCH$_2$CH$_2$, CH$_2$=CHCH$_2$, and CF$_3$CH$_2$CH$_2$). The resulting Li$^+$Me$_2$CuR(CN)$^-$ intermediates then afford the observable Me$_3$CuR$^-$ tetraalkylcuprate anions upon Me/CN exchanges with added MeLi. In contrast, the reactions of LiCuMe$_2$·LiCN with neopentyl iodide and various aryl halides gave rise to halogen-copper exchanges. Concentration- and solvent-dependent studies suggested that lithium tetraalkylcuprates partly form Li$^+$Me$_3$CuR$^-$ contact ion pairs and presumably also triple ions LiMe$_6$Cu$_2$R$_2^-$. According to theoretical calculations, these triple ions consist of two square-planar Me$_3$CuR$^-$ subunits binding to a central Li$^+$ ion. Upon fragmentation in the gas phase, the Me$_3$CuR$^-$ anions undergo reductive elimination, yielding both cross- (MeR) and homo-coupling products (Me$_2$). The branching between these channels showed a marked dependence on the nature of R. The fragmentation of LiMe$_6$Cu$_2$R$_2^-$ also affords both cross- and homo-coupling products, but strongly favors the former. This was rationalized by the preferential interaction of the central Li$^+$ ion with two Me groups of each Me$_3$CuR$^-$ subunit, which thereby block the homo-coupling channel.

Finally, the reactivity of organocuprates in conjugate addition reactions was investigated, with cyano-substituted ethylenes C$_2$H$_{n-4}$(CN)$_n$, n = 1 – 4 as Michael acceptors. In the case of acrylonitrile, n = 1, polymerization was induced, but no reactive intermediates were detected. In contrast, the reaction with fumaronitrile, n = 2, permitted the detection of π-complexes in different aggregation states. The identities of the latter were confirmed by the release of intact fumaronitrile upon their fragmentation in the gas phase. The reactions with 1,1-dicyanoethylene, n = 2, did not halt at the stage of the π-complexes, but proceeded all the way to Michael adducts. In the case of tricyanoethylene, n = 3, dimeric polycyano carbanions were formed. For tetracyanoethylene, n = 4, the reaction instead leads to Cu(III) species, which undergo reductive eliminations. Thus, all intermediates commonly proposed for the conjugate addition of organocuprates to Michael acceptors were detected, providing strong evidence for the currently accepted mechanism.

2 Introduction

2.1. Overview

The late transition metal copper forms organometallic reagents of outstanding importance to organic synthesis[1]. The beginning of organocopper chemistry is marked by the preparation of the highly explosive copper(I) acetylide Cu_2C_2 by Böttger in 1859.[2] In the same year, a reaction between CuCl and Et_2Zn was reported.[3] This reaction did not result in the formation of EtCu, but produced metallic mirrors instead. Therefore, it was concluded that it was impossible to bind an alkyl group to copper. The isolation of phenylcopper from the reaction between CuI and a phenyl Grignard reagent[4] (Reich, 1923) was reported only more than sixty years later. Pioneering work by Gilman in 1936[5] demonstrated the applicability of organocopper reagents in synthetic organic chemistry (Scheme 2.1.1).

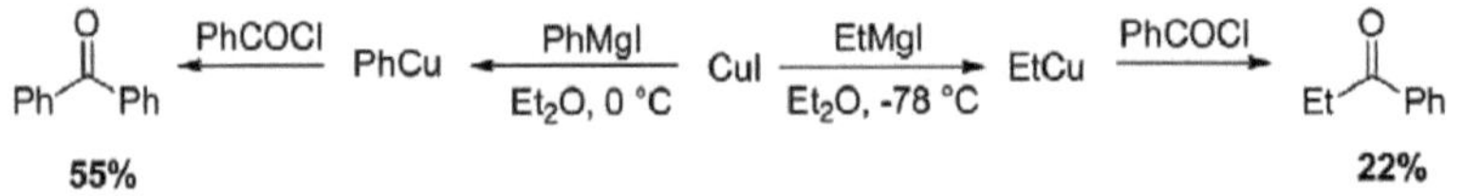

$$Ph\underset{55\%}{\overset{O}{\underset{\|}{C}}}Ph \xleftarrow{PhCOCl} PhCu \xleftarrow[Et_2O, 0\ °C]{PhMgI} CuI \xrightarrow[Et_2O, -78\ °C]{EtMgI} EtCu \xrightarrow{PhCOCl} Et\underset{22\%}{\overset{O}{\underset{\|}{C}}}Ph$$

Scheme 2.1.1. Pioneering investigation of organocopper reactivity by Gilman *et al.*

Moreover, in 1952 the group of Gilman *et al* described the Et_2O-soluble $LiCuMe_2$ reagent,[6] an example of what we now call Gilman cuprates (Scheme 2.1.2). The demonstration of the synthetic potential of these compounds by Corey,[7] House[8,9] and Posner[10] marked a major breakthrough in the field of copper-mediated synthetic organic chemistry.

$$LiCuMe_2 \xleftarrow[Et_2O, -15\ °C]{2\ MeLi} CuI \xrightarrow[Et_2O, -15\ °C]{MeLi} MeCu \xrightarrow{MeLi} $$

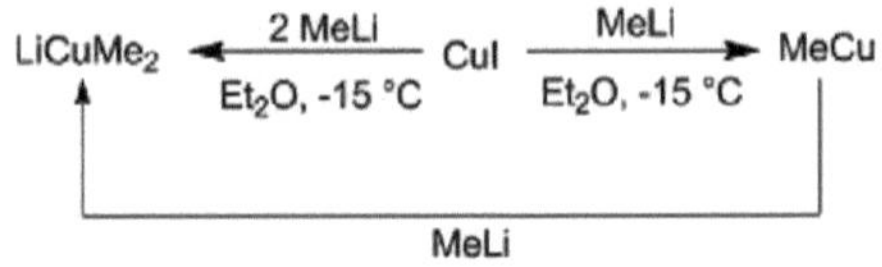

Scheme 2.1.2. Original preparation of the Gilman reagent.

After these studies, a large number of investigations followed, describing the preparation of new types of organocopper reagents and their synthetic applications. Notable examples are the Normant,[11] Lipshutz[12], and Knochel[13] cuprates (Scheme 2.1.3).

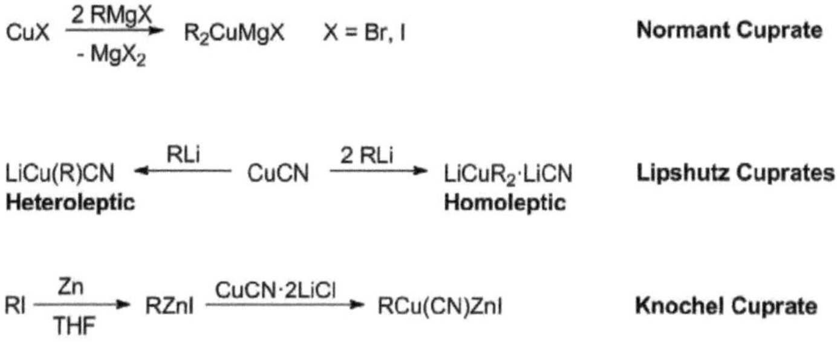

2 Introduction

CuX $\xrightarrow[-\text{MgX}_2]{2\ \text{RMgX}}$ R$_2$CuMgX   X = Br, I   **Normant Cuprate**

LiCu(R)CN $\xleftarrow{\text{RLi}}$ CuCN $\xrightarrow{2\ \text{RLi}}$ LiCuR$_2$·LiCN   **Lipshutz Cuprates**
Heteroleptic   **Homoleptic**

RI $\xrightarrow[\text{THF}]{\text{Zn}}$ RZnI $\xrightarrow{\text{CuCN·2LiCl}}$ RCu(CN)ZnI   **Knochel Cuprate**

Scheme 2.1.3. Preparation of Normant, Lipshutz, and Knochel cuprates.

Among the different variants of organocuprates described above, cyanocuprates are arguably one of the most popular ones. They are easily prepared by transmetallation of CuCN with organolithium reagents RLi, and, depending on the amount of RLi used (Scheme 2.1.3), can be divided into hetero- and homoleptic cuprates, of stoichiometries LiCuR(CN) and LiCuR$_2$·LiCN, respectively. These species find numerous applications in conjugate additions,[14-16] carbocuprations of alkynes,[14a,17] epoxide opening reactions,[14] and nucleophilic substitutions of alkyl halides[14,15,18] and sulfonates (Scheme 2.1.4).[14a]

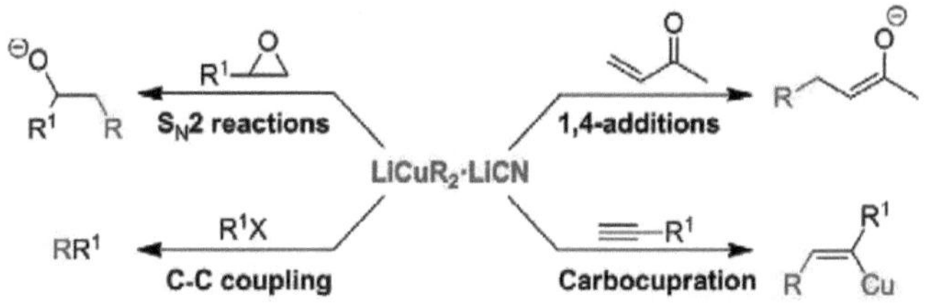

Scheme 2.1.4. Formation and reactivity of lithium cyanocuprates.

Despite having enjoyed tremendous success since their discovery in 1973,[12a] cyanocuprate structure, aggregation and reaction mechanism still have not been fully understood.[19-21] In his review on organocuprate conjugate addition[22], Woodward compares it with a 'black box'. Decoding this 'black box' and shedding light on the often intriguing cyanocuprate chemistry promises to improve the existing synthetic procedures and help devise new ones, allowing us to tap the full synthetic potential of these truly multi-faceted reagents.

2.2. Aggregation and Structure of Cyanocuprates

The high reactivity of cyanocuprates has provoked numerous mechanistic and structural investigations.[19-21] In particular, the βinding site of the cyanide anion has been discussed controversially. Originally, Lipshutz and coworkers postulated the formation of so-called higher-order diorganocuprates $Li_2CuR_2(CN)$, in which the CN^- ions coordinate to the Cu centers.[14,23] Based on ^{13}C NMR and X-ray absorption spectroscopic measurements as well as on theoretical calculations, Bertz[24] and others[25] challenged this view and instead proposed the existence of lower-order diorganocuprates $LiCuR_2·LiCN$. These species resemble traditional Gilman-type cuprates, with CN^- bound to Li^+. X-ray crystallographic studies confirmed the lower-order nature of cyanocuprates,[26,27] which since then has been generally accepted.[28] After the end of this dispute, the question of the aggregation state of cyanocuprates received increasing attention. For solutions of $LiCuR_2·LiCN$ in Et_2O, extensive NMR spectroscopic experiments by Gschwind and collaborators point to the predominance of dimeric contact ion pairs 1 (Scheme 2.2.1),[29] which presumably form even larger, chain-like oligomers.[20,30]

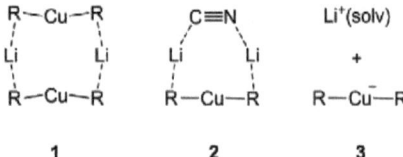

Scheme 2.2.1. Proposed structures of organocopper species present in solutions of cyanocuprates $LiCuR_2·LiCN$ in ethereal solvents (for **1** and **2**, coordinating solvent molecules are omitted for clarity).

In contrast, the situation is less clear for solutions of $LiCuR_2·LiCN$ in THF. IR[31] and X-ray absorption[32] spectroscopic experiments indicate the presence of the contact ion pair **2** in THF (R = Me), which is also consistent with the results of cryoscopic measurements.[33] Moreover, the predominance of **2** was inferred from ^{15}N NMR spectroscopic studies of $LiCu^nBu_2·LiCN$ in THF.[34] However, Gschwind, Boche, and coworkers detected only very small $^1H,^6Li$ HOESY NMR cross signals for $LiCuR_2·LiCN$ in THF (R = Me, CH_2SiMe_3) and thus concluded that these cyanocuprates preferentially form solvent separated ion pairs, i.e., $CuR_2^-/Li(THF)_4^+$ (**3**), in this relatively strongly coordinating solvent; the small cross signals observed were assigned to minor equilibrium populations of the dimeric contact ion pair

1.[20,29a,b] The assumed preponderance of solvent separated ion pairs in THF seems to be in line with X-ray crystallographic data[20,29a,b] and can also rationalize the relative rates of conjugate addition reactions,[24c,35] for which the participation of lithium centers is considered essential.[36] Yet, it apparently is in conflict with the results of the earlier IR, X-ray absorption, cryoscopic, and ^{15}N NMR spectroscopic experiments.

Thus, important aspects of the association/dissociation equilibria of lithium organocuprates still await clarification. In particular, it remains to be shown whether the observed solvent dependence of the equilibrium is a general phenomenon seen for a larger series of lithium organocuprates and other solvents, in addition to THF and Et$_2$O. Moreover, only very little is known about the association/dissociation equilibria of related heteroleptic cuprates, such as LiCuR(CN). This reflects the inherent difficulties of determining the aggregation state of cyanocuprates in solution by spectroscopic methods, which probe this quantity only in a rather indirect manner. As an alternative and possibly more direct approach to identify the nuclearity of cuprate anions, Lipshutz et al. therefore employed ESI mass spectrometry.[37] With this method, a multitude of inorganic[37a] and organometallic cuprate anions could be observed, the latter bearing 2-thiophenyl, alkynyl, and (trimethylsilyl)methyl substituents.[37b] However, analogous experiments probing the more sensitive methyl- and butylcuprate anions were reported to be unsuccessful.[37b] Because of the apparent difficulties in producing such non-stabilized organocuprates by direct ESI, O'Hair and coworkers chose to prepare these species from gaseous precursor ions.[38] In this way, these authors generated many different mononuclear diorganocuprate anions CuR^1(R^2)$^-$ (Scheme 2.2.2) and investigated the gas-phase reactivity of selected examples.

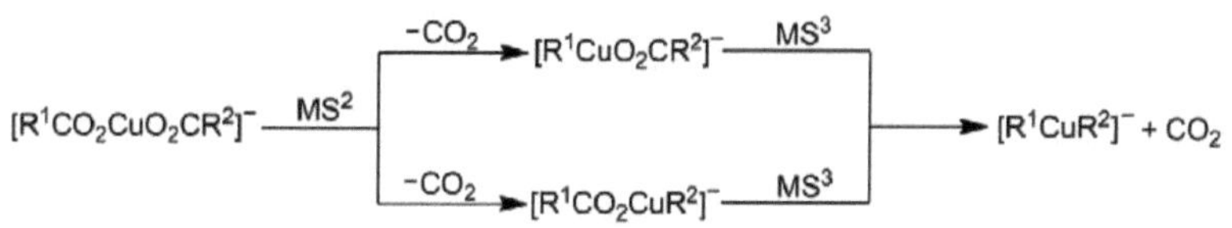

$$[R^1CO_2CuO_2CR^2]^- \xrightarrow{MS^2} \begin{array}{c} \xrightarrow{-CO_2} [R^1CuO_2CR^2]^- \xrightarrow{MS^3} \\ \xrightarrow{-CO_2} [R^1CO_2CuR^2]^- \xrightarrow{MS^3} \end{array} [R^1CuR^2]^- + CO_2$$

Scheme 2.2.2. Generation of mononuclear diorganocuprate anions by gas-phase decarboxylation.

These studies offer detailed insight into both the intrinsic and bimolecular reactivity of organocuprate anions. In contrast to the direct ESI approach pursued by Lipshutz and coworkers, O'Hair's gas-phase preparation does not provide any polynuclear organocuprate ions, and thus cannot be used for investigating aggregation effects.

2.3. Cross-Coupling Reactions

Despite their structural diversity, virtually all of the organocopper reagents known correspond to Cu(I) species with a $3d^{10}$ valence electron configuration,[39,40] like the cyanocuprates described above. However, their reactions with carbon electrophiles, such as alkyl halides, epoxides, and Michael acceptors, have long been postulated to involve $3d^8$ Cu(III) intermediates,[41] which were also predicted by theoretical calculations.[19a,36b,c,42] Because of their supposedly high propensity toward reductive elimination, these copper(III) species were believed to be too elusive for detection.[41,43] Nevertheless, Bertz, Ogle, and coworkers[44] as well as the Gschwind group[45] have recently succeeded in the preparation of several organocopper(III) compounds (Scheme 2.3.1) and their characterization by low-temperature NMR spectroscopy.

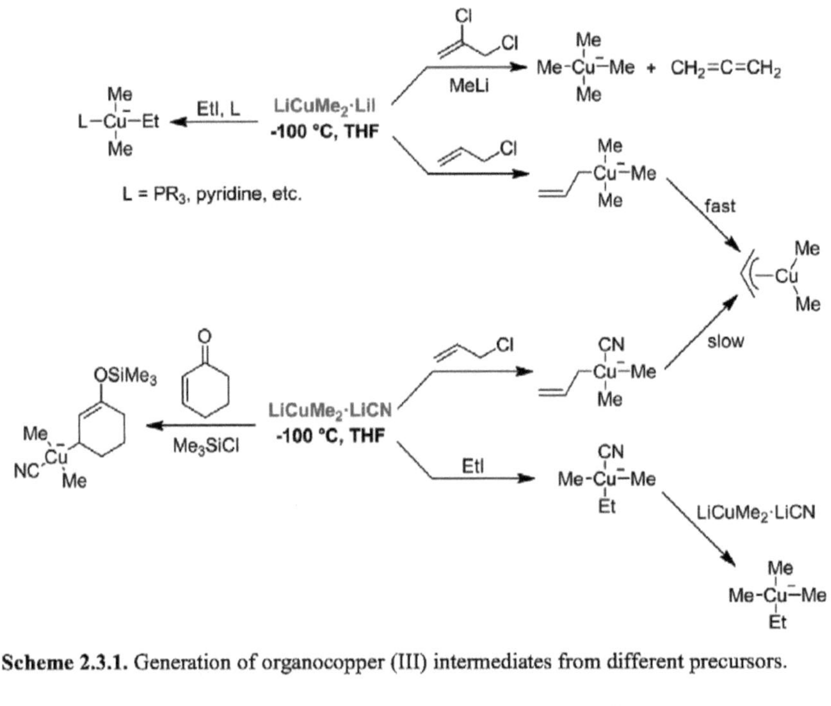

Scheme 2.3.1. Generation of organocopper (III) intermediates from different precursors.

Most of the species detected have a square-planar tetracoordinated Cu(III) core with three alkyl groups of various complexity bound. The nature of the fourth group is variable: neutral donor ligands, alkyl groups or cyanide all help stabilize the otherwise unstable tricoordinate neutral Cu(III) center. The cyanide ligand, if present, can undergo displacement by alkyllithiums, the alkyl groups being better σ-donors than CN^-.[44b,f]

Besides being of fundamental importance, a better understanding of organocopper(III) compounds and their reactivity promises practical benefits, as it might help to optimize reagents and reaction conditions rationally. Among the Cu(III) species so far identified, the tetraalkylcuprate anions are particularly interesting. These species have been observed in the course of cross-coupling reactions between lithium dimethylcuprate $LiCuMe_2 \cdot LiCN$ and alkyl halides RX and in some cases were even found to survive warming-up to 20 °C for short times.[44b,e] This enhanced stability renders tetraalkylcuprates ideal model systems not only for studying the generation of organocopper(III) compounds, but also for probing their reactions.

2.4. Conjugate Addition Reactions

Conjugate additions of organocuprates are one of the most important methods for C–C bond formation, combining broad applicability with stereoselectivity potential.[46] Despite these transformations being so popular, their mechanism is still not entirely understood. Latest investigations, however, helped clarify some mechanistic issues and confirmed the existence of previously assumed intermediates.[15,20a, 22,28,44a-c,g,45,47] Thus, it is currently accepted that the first step of a conjugate addition is the association of the organocuprate with the substrate to form a π-complex, followed by conversion to a Cu(III) species, which then undergoes reductive elimination to yield the anion of the product (Scheme 2.4.1). This, in turn, can be protonated by means of aqueous work-up or be trapped by other electrophiles.

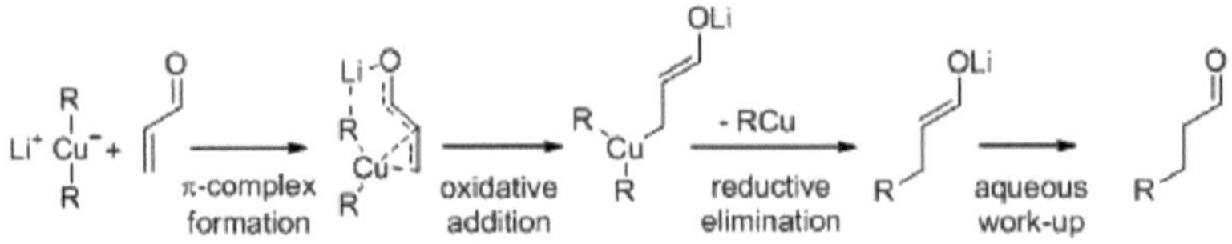

Scheme 2.4.1. Proposed mechanism of reaction between organocuprates and Michael acceptors, exemplified by an α,β-unsaturated carbonyl compound.

Despite the significant number of π-complexes observed by NMR spectroscopy,[48-53] their characterization is by no means comprehensive, and detailed structural investigations are few. Among these are studies of complexes between $LiCuMe_2 \cdot LiX$ (X = I, CN) and 2-cyclohexenones or 10-methyl-$\Delta^{1,9}$-2-octalone[45b], as a result of which some general structural features were determined. Thus, the C=C bond was found to be coordinated by the cuprate moiety, which is bent as a result, whereas the carbonyl group is complexed by lithium.[45b,50b,c] The latter interaction is believed to be important,[50a] both in terms of π-complex stability, and formation of the conjugate addition product. In THF, where monomeric species were detected, the complexing moiety was shown to be either Li or a Li–X–Li salt bridge (X = I, CN).[47a] In contrast, much larger aggregates were observed in the less polar diethyl ether, where the carbonyl group was complexed by both salt and cuprate units.[45b] The exact composition of these aggregates is still unclear. Therefore, further investigations are needed for a deeper mechanistic understanding of organocuprate conjugate additions.

2.5. Objectives

At present, the mechanistic understanding of organocuprate reactions is still far from complete. Evidence on their aggregation state and stoichiometry in solution (probed mainly by NMR spectroscopy and cryoscopy) is rather indirect, and does not account for fluxionality and complex equilibria operative. Reactivity studies performed with these methods thus inevitably suffer from averaging over all species present. To address these issues, the present thesis conducts a systematic investigation of model cyanocuprate systems by ESI-MS, which furnishes direct stoichiometric information, and permits *isolation* and reactivity investigation of all species detected.

For a pilot project, THF solutions of $LiCuR_2 \cdot LiCN$ and $LiCuR(CN)$, (R = Me, Et, ^{n}Bu, ^{s}Bu, ^{t}Bu and Ph) are to be analyzed by ESI-MS, to study the influence of stoichiometry and the nature of R on solution-phase composition. Subsequently, the role of solvent in aggregation/association equilibria is sought to be investigated, by probing cyanocuprate solutions in MeTHF, Et_2O, CPME and MTBE. Complementary electrical conductivity measurements on selected systems should provide additional insight.

Following their detection, the gas-phase reactivities of cyanocuprates, both unimolecular (CID) and bimolecular (reactions with background water) are investigated. On the basis of the results obtained, correlations between aggregation level and reactivity are attempted, together with elucidation of some structural features.

Finally, the synthetically most useful reactions of organocuprates, namely C–C cross-couplings and conjugate additions are focused on. In investigations of chosen model systems, detection of elusive intermediates proposed for these reactions is attempted, and light is shed on their structure and reactivity. These are further investigated with the help of theoretical calculations (performed by Dr. Harald Brand),[54] which allow an even deeper insight. For selected systems, the effect of the solvent on the observed aggregation states is also studied.

3 Instrumentation and Methods

3.1. Electrospray Ionization Mass Spectrometry (ESI-MS)

3.1.1. Theoretical Overview

Mass spectrometry operates by generating gas-phase ions, separating these ions by their mass-to-charge ratio (m/z) and detecting them qualitatively and quantitatively by their respective m/z and abundance. A mass spectrometer always contains the following modules:[55]

- A device to introduce the analyte, e.g. a direct insertion probe or a chromatograph.
- An ionization source, which produces gas-phase ions from the sample.
- One or several analyzers, employing electromagnetic fields, which discriminate between the different ions based on their m/z ratio.
- A detector to register the abundance of the ions emerging from the last analyzer.
- A computer to control the instrument and process the mass spectra.

In this work, two different instruments employing electrospray ionization (ESI) as ionization technique were used: a HCT quadrupole ion trap mass spectrometer (Bruker Daltonik) and a TSQ 7000 multistage mass spectrometer (Thermo-Finnigan).

Electrospray Ionization

ESI is a soft ionization technique, resulting in little or no fragmentation of the ions analyzed.[56] Due to this fact, it has primarily been employed in analysis of multiply charged protein ions.[57] Later on, its use was extended to polymers and different small polar molecules. Very recently, the applicability of ESI in detection and characterization of various organometallic ate complexes has been shown. [58-60]

During the ESI process[61], a weak flux (1 – 10μL min^{-1}) of a dilute analyte solution passes through a capillary tube, to which a high potential is applied (3 – 6 kV). This potential generates an electric field of the order 10^6 V m^{-1}, which induces charge accumulation at the liquid surface located at the tip of the capillary (Scheme 3.1.1.1).

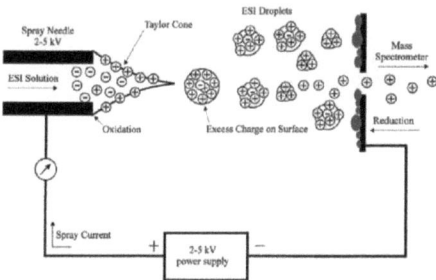

Scheme 3.1.1.1. Schematic of the ionization electrospray process according to ref. 55.

The pressure stemming from this charge accumulation makes the liquid protrude from the capillary. When it becomes higher than the surface tension, the shape of the drop changes to a so-called "Taylor cone", and small charged droplets, containing an excess of charge, are detached. A coaxial stream of inert gas helps limit the dispersion of the resulting electrospray. The droplets formed move in the applied field towards the entrance of the mass spectrometer, passing through a curtain of heated inert gas on their way, and generate ions by one of the two suggested mechanisms.[62,63,64] The first is known as ion evaporation,[62] and assumes that the increased charge density, due to evaporation of the solvent, eventually reaches a value when Coulombic repulsion at the surface of the droplets becomes large enough for desorption of individual ions into the gas phase to occur. According to the second mechanism[63] (charge residue model), the repelling coulombic forces at some point overcome the cohesion forces, causing division of the droplets. They further undergo a cascade of ruptures, yielding smaller and smaller droplets, up to a point when all solvent molecules have evaporated. The current opinion is that small ions form via the ion evaporation mechanism, whereas heavy ions (such as charged proteins) originate according to the charge residue model.[65]

In the ion evaporation mode, the ions detected by ESI do not stem directly from the analyte solution, but rather from the surface of the nanodroplets formed. Therefore, the sensitivity is higher for more surface-active compounds,[56,64] i.e., those analytes present at the surface of the droplets can mask the compounds that are present in the bulk. This phenomenon makes quantitation of obtained results difficult.

3 Instrumentation and Methods

Analyzers.

Once produced, the ions need to be separated according to their m/z ratios. There are a great variety of analyzers, which, however, can be grouped into several major classes:

- Time-of-flight (TOF)
- Quadrupoles (Q)
- Ion traps (Quistors)
- Sector field
- Ion-cyclotron resonance (ICR)

Instruments that have more than one analyzer are called tandem (MS/MS) mass spectrometers and allow structural and sequencing studies to be carried out.[55] The two devices used in this work operate with a tandem quadrupole analyzer, which separates the ions spatially (TSQ 7000 instrument) and with a three-dimensional ion trap, which separates the ions temporally, and allows multistage MS/MS experiments to be carried out (HCT instrument).

Tandem quadrupole analyzer

The TSQ 7000 instrument employs two independent mass analyzers (quadrupoles Q_1 and Q_2) separated by a collision cell (octopole O_1), as depicted in Scheme 3.1.1.2.

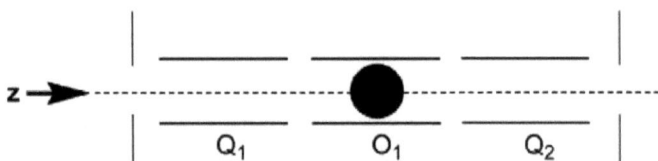

Scheme 3.1.1.2. Diagram of the TSQ 7000 quadrupole analyzer. The two quadrupoles are true mass analyzers, whereas the central octopole is a collision cell made up of an octopole using radio frequency (RF) only.

The quadrupoles consist of four, and the octopole of eight parallel metal rods of alternating polarity. For the quadrupoles, a radio frequency voltage is applied to the rods, with a superimposed direct current voltage. In contrast, an RF-only voltage is applied to the octopole. Ions entering the quadrupole along the z-axis (parallel to the rods) experience forces in the xy-plane perpendicular to it, which cause oscillatory motion. If the oscillations are too large, the ions discharge on the rods before reaching the detector. Since the oscillation amplitude depends on *m/z*, only certain ions can pass through at a given time. This permits selection of ions with particular *m/z* values or allows one to scan across a range of values by continuously varying the applied voltages.

In contrast to both quadrupoles, which act as mass filters, the octopole (O_1) in the middle allows all ions to pass through, due to the absence of a direct current voltage. It can be filled with argon and employed as a collision cell. Instruments with this analyzer setup can be scanned in different ways, with the most important summarized below:

1. Product ion scan. In this mode, an ion with a chosen *m/z* ratio is selected by the first quadrupole. This ion collides with the inert gas atoms inside the central octopole and fragments. The reaction products are analyzed by the second quadrupole.

2. Precursor scan. The second quadrupole is configured to let only a selected ion pass through, whereas the first quadrupole is scanned across a broad mass range. All of the ions that produce the ion with a selected mass are thus detected.

3. Neutral loss scan. In this mode, both quadrupoles are scanned together, with a constant mass offset Δm between them. In this case, only ions of mass *m* that yield fragments with mass $m - \Delta m$ are detected.[55]

Three-Dimensional Ion Trap

The operation principle of quadrupole ion traps is similar to those of standard quadrupole analyzers. The trap itself consists of two conical endcap electrodes, with a donut-shaped ring electrode in between (Scheme 3.1.1.3).

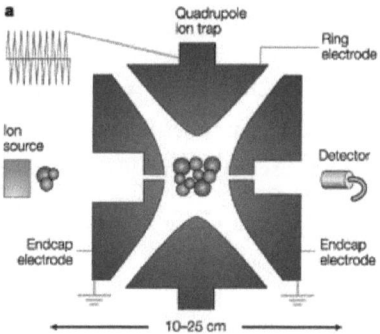

Scheme 3.1.1.3. Schematic representation of the quadrupole ion trap.

Openings at the center of the endcaps allow ions to pass in and out. A high voltage radio frequency (RF) potential is applied to the ring (main potential), whereas the endcaps are held at zero. The oscillating potential forms a substantially quadrupolar field, which can trap ions of a particular mass range. These ions are derived from an external source, and therefore possess a certain amount of kinetic energy. In order for them to be captured in the potential well created by the trap, some of their kinetic energy needs to be dissipated. To achieve this, a collision gas (most usually He) is introduced into the trap to extract energy from the ion beam and cause retention of a certain portion of ions entering the trap.

When trapped, the ions undergo periodic motions in radial and axial directions. The axial oscillations (in the direction of the endcaps) have a certain secular frequency f_z, which is a function of the ion m/z ratio, the RF frequency v, and the RF amplitude V of the main potential. To eject the ions, an auxiliary RF potential is applied to the endcaps, and the amplitude of the main potential V is progressively increased. In doing so, ions with different m/z values (and hence secular frequencies f_z) are brought in resonance with the applied auxiliary frequency, take up energy and are ejected from the trap towards the detector in the axial direction. At the end of the scan, the main potential is dropped to zero to remove the remaining ions from the trap.

For tandem mass spectrometry, a broadband composite of frequencies is applied to the endcaps, ejecting all of the ions stored by resonance, except for the precursor ion. To induce fragmentation, it is then brought into resonance by the auxiliary field, whose amplitude (V_{exc}) in this case is lower than the one used for ejection. The resonating precursor ion takes up energy and begins to collide with the He gas, which causes fragmentations. The product ions, together with the parent ion, are scanned out by resonant excitation as described above. In comparison to a quadrupole instrument, fragmentations in an ion trap are mass-selective. Little energy remains in the product ions to result in subsequent fragmentation. Moreover, these ions do not continue to be excited, because they are not resonated by the auxiliary frequency.[55,66]

3.1.2. Potential and Limitations of ESI-MS

Compared to conventional analysis methods of fluxional species (NMR, UV/Vis and IR spectroscopy, cryoscopy, X-ray crystallographic analysis), ESI-MS offers significant advantages. So, ion exact mass, together with its isotope pattern provide unambiguous stoichiometric information, which can be further supplemented by fragmentation MS/MS experiments. The possibility of isolating ions of interest also provides a unique possibility to study the gas-phase reactivities of all ionic species present separately, which, of course, is not possible in the condensed phase. Moreover, ESI-MS is insensitive towards equilibrium averaging effects, unlike NMR, thus allowing constituents of complex mixtures to be identified and characterized. When compared to other MS ionization techniques, ESI imparts the lowest amount of energy to the generated ions, allowing studies of weak non-covalent interactions to be carried out. In particular, the formation of triple ions AB_2^- from contact ion pairs A^+B^- and free ions B^- in solution has been recently characterized by ESI-MS, [59,60,67] together with ion association in general.[68d]

On the downside, just like any other MS technique, only charged species can be detected by ESI-MS. Moreover, quantitative analysis is not trivial, due to several factors. First, the detected analyte ions do not stem directly from the sampled solution, but rather from charged nanodroplets generated in the course of the ESI process.[62] Previous studies have shown that the analyte concentration in these nanodroplets is higher than in the sampled solutions[62] and that their effective temperature may also change,[69] in other words, the composition of nanodroplets can be different from that of the probed solution. The increased analyte concentration can, for instance, result in a shift to higher aggregation states, in accord with the law of mass action. Second, different analytes have different response factors, i.e., tendencies to evaporate into the gas phase from the nanodroplet surface. These response factors are correlated with the surface activity of the analyte:[56] the one most surface active has a higher tendency to be ejected into the gas phase and be detected. Another disadvantage of ESI-MS is the possible formation of neutral molecule adducts with ions. Beneficial in the analysis of polar non-ionizable species, this ESI artifact can be detrimental to analysis of organometallic aggregation states, producing ions that do not exist in solution.

Altogether, ESI-MS is a useful tool to probe the qualitative speciation of ions in solution. For quantitative results, more involved experiments are necessary, together with coupling to classical analytical methods.

3.1.3. Experimental Part
Settings used for the TSQ 7000 instrument

Sample solutions ($c \approx 10 - 25$ mM) were transferred into a gas-tight syringe and introduced into the ESI source of the instrument at flow rates of ca. $0.6 - 3.0$ mL h^{-1} by means of a syringe pump. Particular care was taken to exclude or minimize contact of the organometallic samples with air. Traces of moisture or oxygen in the inlet system were eliminated by extensively flushing it with dry THF before adding the organometallic sample. The sample solution entered the source via a fused silica tube (0.10 mm inner diameter). Stable electrospray conditions were achieved for ESI voltages of ± 3.5 kV with nitrogen as sheath gas (2.5 bar). The electrospray then passed a heated capillary, which was held at temperatures from 60 to 150 °C. The potential difference between the capillary and the ion optic lenses was kept low to avoid strong acceleration of the ions and unwanted fragmentations due to energetic collisions with gas molecules present in the ESI source region.

For probing the unimolecular gas-phase reactivity, argon (Linde, 99.998%) was used as collision gas in the octopole ($p(\text{Ar}) \approx 0.6$ mTorr). The vacuum chamber of the mass spectrometer was held at $T \approx 343$ K, and it is assumed that this temperature also describes the distribution of the internal energy of the neutral reactants/the collision gas.

Settings used for the HCT instrument

Sample solutions of $c = 25 - 100$ mM were administered into the ESI source via a syringe pump at flow rates of $1 - 4$ mL h^{-1}. With these settings, hydrolysis and/or oxidation reactions could be suppressed almost completely, whereas products of such degradation reactions were observed for samples of lower concentrations administered at lower flow rates. The source of the HCT ion trap was operated with N_2 as sheath gas (0.7 bar backing pressure), an ESI voltage of ± 3 kV, and N_2 as drying gas (5 L min^{-1}). The latter was held at 60 °C in order to minimize thermal decomposition, although higher temperatures did not show such decomposition reactions for $LiCu^nBu_2 \cdot LiCN$ sample solutions. The thus produced ions then passed a capillary, a skimmer, and two transfer octopoles before entering the quadrupole ion trap. Varying the voltage offsets of the capillary exit (Figure 3.1.3.1) and the transfer octopoles (Figure 3.1.3.2) had significant effects. For higher absolute voltages, the ratio $I(Li_2Cu_3{}^nBu_6{}^-)/I(Cu^nBu_2{}^-)$ strongly decreased because of fragmentation reactions due to energetic collisions with residual gas, as was proven by deliberate fragmentation of mass-selected $Li_2Cu_3{}^nBu_6{}^-$ (see Section 4.2). To avoid these unwanted decomposition reactions, low absolute voltages (V(capillary exit) = ±20 V, V(skimmer) = ±20 V, V(Oct 1 DC) = ±5 V, V(Oct 2 DC) = ±1.7 V) were applied consistently.

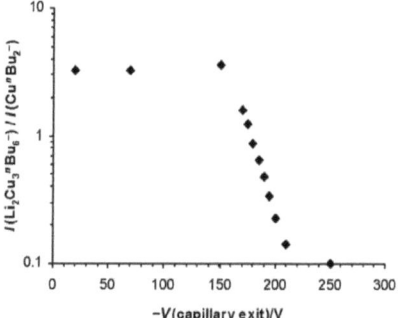

Figure 3.1.3.1. Ratio of the signal intensities of $Li_2Cu_3{}^nBu_6{}^-$ and $Cu^nBu_2{}^-$ produced by ESI of a 25 mM solution of $LiCu^nBu_2 \cdot LiCN$ in THF as a function of the voltage of the capillary exit (other parameters: V(Oct 1 DC) = –5 V, V(Oct 2 DC) = –1.7 V, trap drive level of 20).

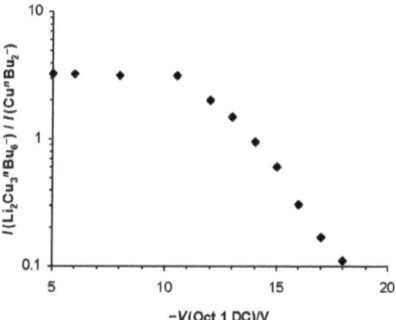

Figure 3.1.3.2. Ratio of the signal intensities of $Li_2Cu_3{}^nBu_6{}^-$ and $Cu^nBu_2{}^-$ produced by ESI of a 25 mM solution of $LiCu^nBu_2 \cdot LiCN$ in THF as a function of the voltage of the first transfer octopole (other parameters: V(capillary exit) = –20 V, V(Oct 2 DC) = –1.7 V, trap drive level of 20).

The quadrupole ion trap itself was filled with helium (Air Liquide, 99.999% purity, estimated pressure $p(\text{He}) \approx 2$ mTorr) and operated at a trap drive level of 20. This low value was chosen on purpose to avoid unwanted fragmentation reactions resulting from too high a kinetic excitation of the trapped ions. At the same time, the trap drive level also affects the relative efficiency of ion ejection toward the detector and thereby discriminates against either light or heavy ions (Figure 3.1.3.3). While the constant trap drive level applied in all experiments ensures the comparability of relative signal intensities for different experiments, it is obvious that no rigorous quantitation independent of mass discrimination is possible.

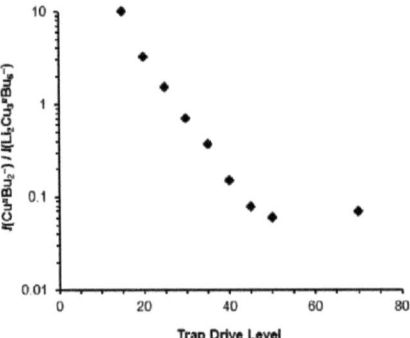

Figure 3.1.3.3. Ratio of the signal intensities of Cu^nBu_2^- and $\text{Li}_2\text{Cu}_3{}^n\text{Bu}_6^-$ produced by ESI of a 25 mM solution of $\text{LiCu}^n\text{Bu}_2 \cdot \text{LiCN}$ in THF as a function of the trap drive level (other parameters: $V(\text{capillary exit}) = -20$ V, $V(\text{Oct 1 DC}) = -16$ V, $V(\text{Oct 2 DC}) = -1.7$ V).

The ions observed were identified based on their m/z ratios, their isotope patterns (see, e.g., Figure 3.1.3.4), and their fragmentation behavior (see Results and Discussion). Typically, m/z ranges of 50 – 1000 were scanned.

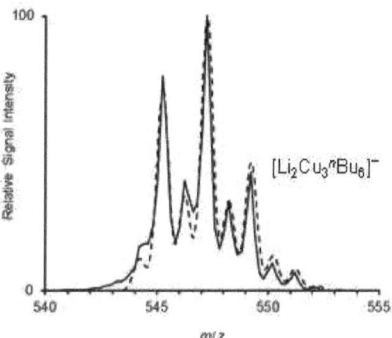

Figure 3.1.3.4. Comparison of observed (plain) and simulated (dashed) isotope patterns for $Li_2Cu_3{}^nBu_6{}^-$.

For gas-phase fragmentation experiments, ions were mass-selected with mass windows of 1 – 2 amu, subjected to excitation voltages of amplitudes V_{exc}, and allowed to collide with He gas. Note that the low-mass cut-off of the ion trap prohibits the detection of fragment ions whose m/z ratio is $\leq 27\%$ of the parent ion.

Comparison of the performance of the TSQ 7000 with the HCT instrument

Preliminary experiments compared the performance of the TSQ 7000 instrument with that of an HCT quadrupole ion trap. While the latter showed clear differences between solutions of $LiCuR_2 \cdot LiCN$ and $LiCuR(CN)$, $R = {}^nBu$ and Ph (Figures 3.1.3.5 and 3.1.3.6), the former did not and invariantly produced ions with R/Cu ratios ≤ 1 (Figures 3.1.3.7 and 3.1.3.8).

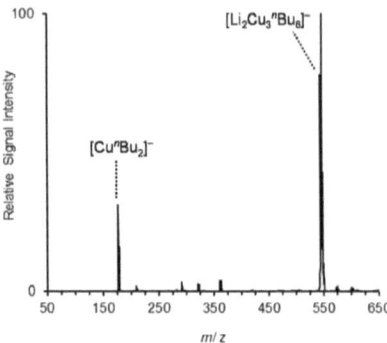

Figure 3.1.3.5. Negative-ion mode ESI mass spectrum of a 25 mM solution of LiCu^{n}Bu$_2$·LiCN in THF, measured by the HCT instrument.

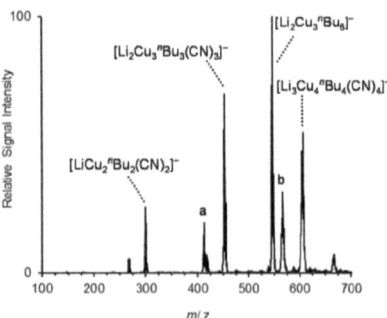

Figure 3.1.3.6. Negative-ion mode ESI mass spectrum of a 25 mM solution of LiCu^{n}Bu(CN) in THF, measured by the HCT instrument; a = Li$_2$Cu$_3$^{n}Bu$_2$(OH)(CN)$_3^-$, b = Li$_3$Cu$_4$^{n}Bu$_3$(OH)(CN)$_4^-$.

Apparently, the experiments with the TSQ 7000 instrument suffered from the occurrence of hydrolysis and/or oxidation reactions, which presumably resulted from an imperfect insulation of the spray from the ambient atmosphere (the predominance of Cu$_2$R$_2$(CN)$^-$ and the deficiency of ions in higher aggregation states furthermore point to fragmentation during the ESI process). Therefore, all further experiments employed the HCT ion trap.

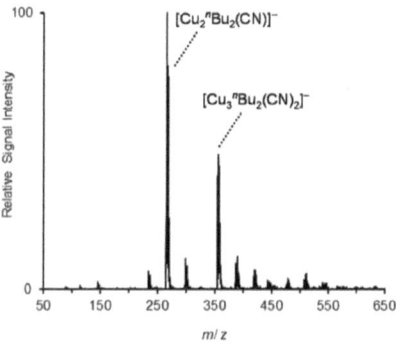

Figure 3.1.3.7. Negative-ion mode ESI mass spectrum of a 25 mM solution of LiCu^{n}Bu$_2$·LiCN in THF, measured with the TSQ instrument.

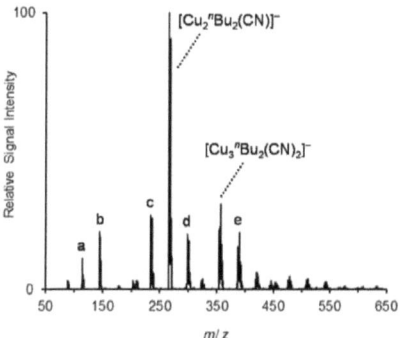

Figure 3.1.3.8. Negative-ion mode ESI mass spectrum of a 25 mM solution of LiCu^{n}Bu(CN) in THF, measured with the TSQ instrument, a = Cu(CN)$_2^-$, b = Cu^{n}Bu(CN)$^-$, c = Cu$_2^n$Bu(CN)$_2^-$, d = LiCu$_2^n$Bu$_2$(CN)$_2^-$, e = LiCu$_3^n$Bu$_2$(CN)$_3^-$.

3.1.4. Analysis of Energy-Dependent Fragmentation Reactions

To investigate whether excitation voltages V_{exc} of the gas-phase fragmentation reactions could be converted into absolute energies in a straightforward manner,[70] the dissociation behavior of a series of benzylpyridinium ions (Scheme 3.1.4.1, Figures 3.1.4.1 and 3.1.4.2 and Table 3.1.4.1) was studied; the activation energies associated with their dissociation had previously been derived from theoretical calculations.[71]

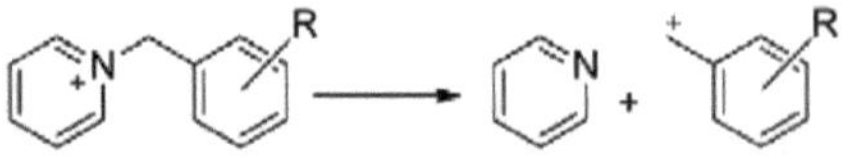

Scheme 3.1.4.1. Collision-induced dissociation of mass-selected benzylpyridinium cations ($R-C_6H_4-CH_2-NC_5H_5^+$).

Unlike the case of Zins et al.,[70b] no satisfactory correlation was found between the obtained appearance voltages V_{appear} of the fragment ions (for the definition of V_{appear}, see Figure 3.1.4.1) and the calculated activation energies AE_{calc} reported in the literature (Figure 3.1.4.2).[70] Hence, a conversion of the V_{exc} values into absolute energies does not appear possible for the employed ion trap.

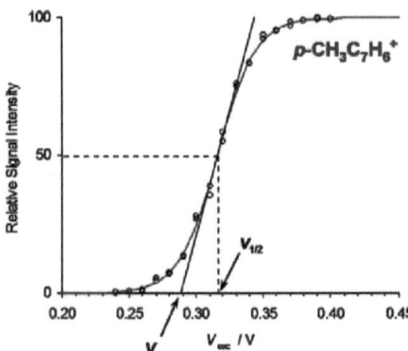

Figure 3.1.4.1. Fragment yield upon collision-induced dissociation of mass-selected p-CH_3-$C_6H_4-CH_2-NC_5H_5^+$ as function of V_{exc}. $V_{1/2}$ corresponds to V_{exc} at the turning point of the sigmoidal fit (50% dissociation of the parent ion). The appearance voltage V_{appear} is given by the x-axis intercept of the tangent line at the point of inflection. For further details, see reference 70b.

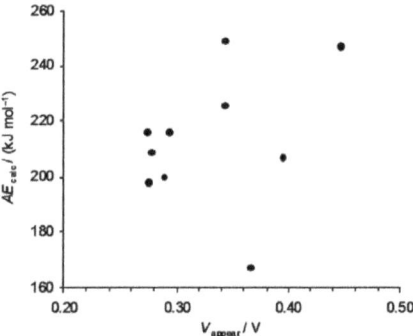

Figure 3.1.4.2. Correlation of experimentally determined appearance voltages V_{appear} with the theoretically calculated appearance energies AE_{calc} of selected benzylpyridinium cations.

Table 3.1.4.1. Experimentally determined appearance voltages[a] and calculated appearance energies[70b] of selected benzylpyridinium cations $[R-C_6H_4-CH_2-NC_5H_5]^+$.

Substitution	V_{appear} / V	AE_{calc} / [kJ·mol^{-1}]
H	0.314 ± 0.006	226
o-CH$_3$	0.277 ± 0.006	209
m-CH$_3$	0.293 ± 0.007	216
p-CH$_3$	0.288 ± 0.006	200
3,5-Dimethyl	0.394 ± 0.006	207
p-F	0.273 ± 0.008	216
p-I	0.275 ± 0.006	198
p-OCH$_3$	0.365 ± 0.006	167
p-CN	0.446 ± 0.01	247
p-CF$_3$	0.342 ± 0.006	249

[a] Error bars estimated from various fits with sigmoid functions that are still in reasonable agreement with experimental data.

3.2. Electrical Conductivity Measurements

Electrical conductivity measurements were performed with a SevenMulti instrument (Mettler Toledo) and a stainless steel electrode cell (InLab741, Mettler Toledo, κ_{cell} = 0.1 cm^{-1}) calibrated against a 0.1 M solution of aqueous KCl at 298 K. Test measurements of solutions of NaBPh$_4$ in THF showed that the instrument also worked correctly at low temperatures (Figure 3.2.1).[72]

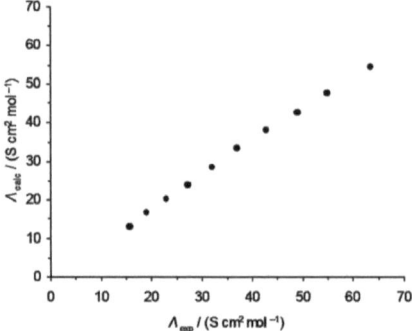

Figure 3.2.1. Comparison of the measured molar conductivity Λ_{exp} and calculated molar conductivity Λ_{calc} of a 80 µM solution of NaBPh$_4$ in THF at temperatures ranging from 203 K to 298 K.

The Λ_{calc} values were calculated from tabulated limiting molar conductivity values Λ_0 and dissociation constants K_{diss} taken from reference 72. Ostwald's dilution law was used to model the dissociation/association processes. The slight deviations observed are likely due to the fact that substance concentrations and not activities were used in the calculations. With increasing electrolyte concentration, the observed discrepancies become more pronounced, as is expected.

3.3. Theoretical Calculations on Tetraalkylcuprates(III)

Theoretical calculations were performed by Dr. Harald Brand[54] with the program package Gaussian 03.[73] All calculations refer to the gas phase, thus making possible a direct comparison with the gas-phase experiments. Similarly to related previous work,[45c,74] a first set of density functional theory (DFT) calculations employed the B3LYP hybrid functional[75] and an effective core potential for the Cu atoms (B3LYP/6-31G*/SDD).[76] As discussed in Section 4.3, the resulting activation energies for the fragmentation reactions of the Me_3CuR^- complexes appeared to be biased in favor of the homo-coupling channel. For the fragmentation of Me_3CuEt^-, exploratory calculations with other methods were therefore performed, including B3LYP/6-31G* all-electron calculations (Table 3.3.1).[77] With a larger basis set and the MDF effective core potential,[78] Møller-Plesset perturbation[79] theory (MP2/6-311+G*/MDF) exhibited a somewhat improved behavior at affordable costs (Table 3.3.1) and was used for further calculations on the mononuclear Me_3CuR^- anions and their unimolecular reactions.

Table 3.3.1. Energies of the Transition States of the Cross-Coupling and Homo-Coupling Reactions for Me_3CuEt^- Calculated with Different Theoretical Methods (Eq 4.3.3.1 and 4.3.3.2).

Theoretical Method	$\Delta_{act}E$ (Eq 4.3.3.1) [kJ mol^{-1}]	$\Delta_{act}E$ (Eq 4.3.3.2) [kJ mol^{-1}]
B3LYP/6-31G*/SDD	148.9	141.4
B3LYP/6-31G*/MDF	148.8	141.5
B3LYP/6-31G* (all electron)	185.7	176.7
B3LYP/6-311+G*/MDF	138.9	135.2
MP2/6-31G*/SDD	132.6	127.9
MP2/6-31G*/MDF	132.6	127.9
MP2/6-311+G*/MDF	131.1	131.4
MP2/6-311+G**/MDF	131.6	130.7
MP2/6-311++G**/MDF	132.5	131.3
MP2/cc-pVTZ/MDF	136.9	131.8
MP2/GTMP2Large/MDF	136.0	133.6
MP4D/6-311+G*/MDF	129.7	130.6

Vibrational analyses were performed to classify stationary points as local minima (zero imaginary frequencies) or transition states (one imaginary frequency). All energies given are zero-point corrected. Minimum energy structures were calculated for different coordination modes (for the triple ions LiMe$_6$Cu$_2$R$_2^-$), but not the complete conformational space was searched. Instead, staggered alkyl chain conformations were used as starting points for the geometry optimizations. For the case of LiCu$_2$Me$_8^-$, not only B3LYP calculations (B3LYP/6-31G*/SDD), were performed, but other functionals, such as the B3PW91[80] and MPW1PW91 functional,[81] as well as Møller-Plesset perturbation theory were also employed to check the robustness of the predicted coordination geometry. The C-Li and Cu-Li interactions of the resulting optimized structures were also characterized by natural bond orbital analyses (Table 3.3.2).[82]

Table 3.3.2. Atom-atom overlap-weighted natural atomic orbital (NAO) bond orders of selected bonds in [LiCu$_2$Me$_8$]$^-$ derived from natural bond order (NBO) analyses for various different theoretical methods.

	Li-C$_{Me1}$	Li-C$_{Me2}$	Li-C$_{Me3}$	Li-C$_{Me4}$	Li-Cu1	Li-Cu2
B3LYP/6-31G* all electron	0.1169	0.1222	0.1170	0.1221	0.0814	0.0814
B3LYP/6-31G*/SDD	0.1230	0.1188	0.1230	0.1187	0.0639	0.0640
B3PW91/6-31G*/SDD	0.1200	0.1143	0.1192	0.1147	0.0647	0.0664
MPW1PW91/6-31G*/SDD	0.1200	0.1183	0.1201	0.1181	0.0615	0.0613
HF/6-31G*/SDD	0.1005	0.0985	0.1011	0.0980	0.0450	0.0453
MP2/6-31G*/SDD	0.1134	0.1128	0.1134	0.1127	0.0676	0.0676

Moreover, for the allyl-containing cuprate ions Me$_3$CuR$^-$ and MeCuR$^-$ (R = allyl), not only σ-bound, but also π-bound isomers were considered. The latter were consistently found to be unstable. The DFT method also predicted the transition structure associated with the reductive elimination of MeR from Me$_3$CuR$^-$ to correspond to a σ-bound complex. In contrast, MP2 calculations did not find an analogous σ-bound transition structure, but only a π-bound isomer. To compute nonetheless at least an approximate activation energy for the MeR elimination with this method, a transition structure was considered, with optimized geometry except for the distance between the β-C atom and the Cu center, which was held constant at 250 pm, i.e., the distance derived from the DFT calculations.

4 Results and Discussion

4.1. Aggregation and Structure of Cyanocuprates

In this chapter, the results of ESI-MS and electrical conductivity studies of cyanocuprates are presented, followed by a discussion section, where general trends, effect of solvents and substituents, and other parameters are focused on.

4.1.1. Negative-Ion Mode ESI Mass Spectrometry

LiCuR$_2$·LiCN Solutions in THF. The negative-ion mode ESI mass spectra obtained for solutions of LiCuR$_2$·LiCN in THF (R = Me, Et, ^{n}Bu, ^{s}Bu, ^{t}Bu, Ph) are almost completely dominated by organocuprate anions of the homologous series Li$_{n-1}$Cu$_n$R$_{2n}^-$, $n = 1 - 3$, as illustrated for R = Me (Figure 4.1.1.1). In this case, the di- and trimeric members of the series are both observed in high relative abundance whereas monomeric CuMe$_2^-$ is absent. Ions of smaller signal intensities centered at m/z = 397 (Figure 4.1.1.1 a) and 603 (Figure 4.1.1.1 b) are assigned to higher aggregates Li$_{n-1}$Cu$_n$Me$_{2n}^-$, n = 4 and 6, respectively, in which the methyl substituents are partially exchanged for hydroxyl groups. Here, hydrolysis reactions are apparently not suppressed completely.

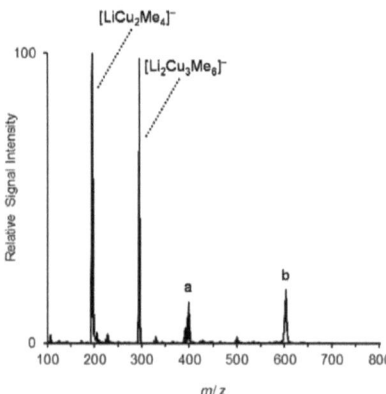

Figure 4.1.1.1. Negative-ion mode ESI mass spectrum of a 25 mM solution of LiCuMe$_2$·LiCN in THF, a = Li$_3$Cu$_4$Me$_{8-x}$(OH)$_x^-$, b = Li$_5$Cu$_6$Me$_{12-x}$(OH)$_x^-$, x = 1 – 3.

With the notable exception of the ^{t}Bu system, all of the other LiCuR$_2$·LiCN solutions also show the trimeric complex in high relative signal intensity (Table 4.1.1.1). In contrast, the dimeric cuprate is considerably less abundant. For R = ^{n}Bu and, in particular, R = ^{s}Bu, ^{t}Bu, and Ph (Figure 4.1.1.2), the CuR$_2^-$ monomer is also observed in high signal intensity. This finding proves that the relative depletion of the dimeric complex does not result from a mass discrimination effect (see Section 3.1.3) but rather reflects its intrinsically lower tendency of formation. The absence of any hydroxyl-containing ions, apart from those in Figure 4.1.1.1, indicates the complete exclusion of hydrolysis reactions.

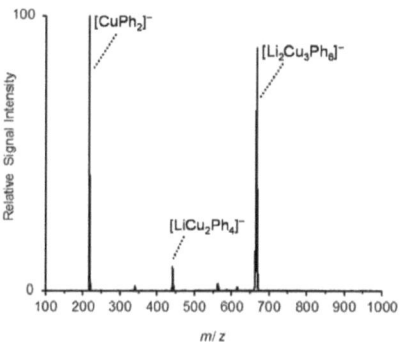

Figure 4.1.1.2. Negative-ion mode ESI mass spectrum of a 25 mM solution of LiCuPh$_2$·LiCN in THF.

Table 4.1.1.1. Organocuprate anions observed upon ESI of THF solutions of LiCuR$_2$·LiCN in high (++) and medium (+) relative abundance.

entry		n	R = Me	Et	^{n}Bu	^{s}Bu	^{t}Bu	Ph
1	Li$_{n-1}$Cu$_n$R$_{2n}^-$	1			+	++	++	++
2		2	++			+		
3		3	++	++	++	++		++

LiCu(Me)R·LiCN Solutions in THF. Besides homoleptic cuprates, mixed cuprates LiCu(Me)R·LiCN, prepared by transmetalation of CuCN with a 1:1 mixture of MeLi and RLi, were also probed. The anions observed all belong to the Li$_{n-1}$Me$_{2n-x}$Cu$_n$R$_x^-$ homologous series (Table 4.1.1.2). With the exception of R = ^{t}Bu, no dimeric but only mono- and trimeric complexes exhibit significant abundance. This trend matches the behavior of the homoleptic $Li_{n-1}Cu_nR_{2n}^-$ anions (R ≠ Me), thus indicating that the larger organyl group R and not the methyl substituent controls the aggregation state of the complexes.

Table 4.1.1.2. Organocuprate anions observed upon ESI of THF solutions of LiCu(Me)R·LiCN in high (++) and medium (+) relative abundance.

entry		n	x	R = Et	^nBu	^sBu	^tBu	Ph
1	$Li_{n-1}Me_{2n-x}Cu_nR_x^-$	1	2		++	++	++	++
2		2	3				+	
3			4				+	
4		3	1					+
5			2					++
6			3	++	+	+		+
7			4	++	++	+		+
8			5	+	++	+		
9			6	+				

The apparently different influence of the methyl and the other organyl substituents on the aggregation state is paralleled by their asymmetric distribution in the detected cuprate anions. As illustrated for the case of R = ^nBu, the observed complexes are enriched in the larger organyl and depleted in the methyl substituent relative to a completely statistical partitioning (Figure 4.1.1.3). This trend holds for all other detected species except for $Li_2Me_{6-x}Cu_3Ph_x^-$, $x = 1 - 6$, for which the methyl-rich anions predominate.

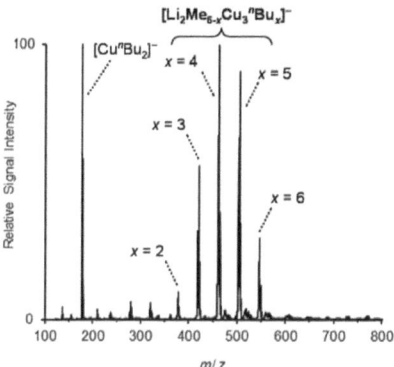

Figure 4.1.1.3. Negative-ion mode ESI mass spectrum of a 25 mM solution of LiCu(Me)^{n}Bu·LiCN in THF.

Virtually identical results are obtained when the mixed samples are prepared by combining two separate solutions of LiCuMe$_2$·LiCN and LiCuR$_2$·LiCN immediately ($\Delta t \approx 2$ min) before the measurement. This result implies that the exchange and equilibration of the different organyl groups occurs relatively fast.

Li$_m$CuR$_m$(CN) Solutions in THF, $m \leq 1$. Anion-mode ESI of solutions of LiCuR(CN) in THF produces richer mass spectra than found for their LiCuR$_2$·LiCN counterparts, as illustrated for the case of R = Me (Figure 4.1.1.4). The majority of observed species can be assigned to the homologous series Li$_{n-1}$Cu$_n$Me$_n$(CN)$_n^-$, $n = 2 - 6$. Note that the stoichiometry of these complexes reflects the nominal overall composition of the sampled solution. The only anion of significant abundance that does not fit into this series corresponds to Li$_2$Cu$_3$Me$_6^-$. This species belongs to the Li$_{n-1}$Cu$_n$R$_{2n}^-$ series already known from the LiCuR$_2$·LiCN samples (see above). As will be discussed in Section 4.1.5, species such as LiCuR$_2$ might possibly form from LiCuR(CN) in Schlenk-type equilibria (along with LiCu(CN)$_2$).

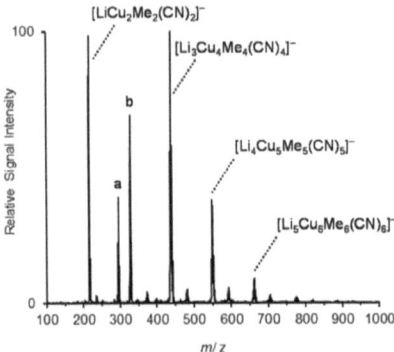

Figure 4.1.1.4. Negative-ion mode ESI mass spectrum of a 25 mM solution of LiCuMe(CN) in THF, a = $Li_2Cu_3Me_6^-$, b = $Li_2Cu_3Me_3(CN)_3^-$.

For all of the other LiCuR(CN) solutions, members of the $Li_{n-1}Cu_nR_n(CN)_n^-$ series are detected upon negative-ion mode ESI mass spectrometric analysis (Table 4.1.1.3 and Figure 4.1.1.5). In no case is the monomer $CuR(CN)^-$ observed, while the relative abundances of the higher aggregates depend on the organyl groups R. In addition, the mass spectra again show the presence of anions of the $Li_{n-1}Cu_nR_{2n}^-$ series. The remaining species of significant intensity correspond to $Cu_2Et_2(CN)^-$ and $Li_{n-1}Cu_nR_{n-1}(OH)(CN)_n^-$, with R = ^{n}Bu and ^{s}Bu and n = 3 and 4. Being particularly prominent in the case of R = ^{s}Bu (Figure 4.1.1.5), the $Li_{n-1}Cu_nR_{n-1}(OH)(CN)_n^-$ complexes stand out from the other cuprate anions in that they contain incorporated hydroxyl ligands. These species were consistently observed in significant abundance for different batches of RLi reagents used in the preparation of the sample solutions. To obtain a definitive proof of identity, isotopic labeling was used in additional experiments on solutions of $Li_{0.8}CuR_{0.8}(^{13}CN)$, R = ^{n}Bu and ^{s}Bu, in THF. Negative-ion mode ESI mass-spectrometric analysis showed the presence of ions at m/z = 415/417/419 and 569/571/573, respectively. In comparison to the corresponding unlabeled ions, the labeled ions are shifted by Δm = 3 and 4, respectively. This means that the complexes in question must contain 3 and 4 cyanide moieties, in accord with their proposed identities as $Li_2Cu_3R_2(OH)(CN)_3^-$ and $Li_3Cu_4R_3(OH)(CN)_4^-$.

Table 4.1.1.3. Organocuprate anions observed upon ESI of THF solutions of LiCuR(CN) in high (++) and medium (+) relative abundance.

entry		n	R = Me	Et	^{n}Bu	^{s}Bu	^{t}Bu	Ph
1	Li$_{n-1}$Cu$_n$R$_n$(CN)$_n^-$	2	++	++	+	+		++
2		3	+	+	+		+	+
3		4	++	+			+	
4		5	+					
5	Cu$_2$R$_2$(CN)$^-$			+				
6	Li$_{n-1}$Cu$_n$R$_{2n}^-$	1				+	++	+
7		2					+	
8		3	+	+	++	++		
9	Li$_{n-1}$Cu$_n$R$_{n-1}$(OH)(CN)$_n^-$	3			+	+		
10		4			+	++		

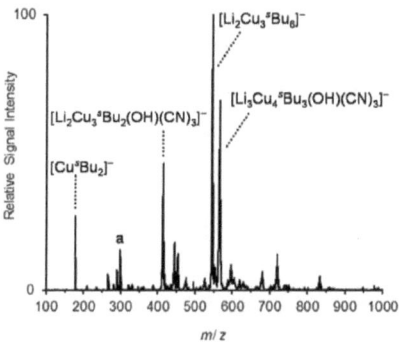

Figure 4.1.1.5. Negative-ion mode ESI mass spectrum of a 25 mM solution of LiCu^{s}Bu(CN) in THF, a = LiCu$_2$^{s}Bu$_2$(CN)$_2^-$.

To rationalize the formation of hydroxyl-containing ions seen in the spectra of LiCuR(CN), R = ^{n}Bu, ^{s}Bu, the possibility that the latter can result from unwanted gas-phase reactions of

4 Results and Discussion

$Li_2Cu_3R_3(CN)_3^-$ and $Li_3Cu_4R_4(CN)_4^-$ with water, Eq (4.1.1.1) and (4.1.1.2), respectively, was considered.

$$Li_2Cu_3R_3(CN)_3^- + H_2O \rightarrow Li_2Cu_3R_2(OH)(CN)_3^- + RH \quad (4.1.1.1)$$

$$Li_3Cu_4R_4(CN)_4^- + H_2O \rightarrow Li_3Cu_4R_3(OH)(CN)_4^- + RH \quad (4.1.1.2)$$

Alternatively, hydrolysis might arise from residual traces of moisture/air in the inlet system. However, the brief exposure of a solution of $LiCu^sBu_2 \cdot LiCN$ to air only led to the detection of $Cu_2{}^sBu_2(CN)^-$, $Cu_3{}^sBu_2(CN)_2^-$, and $Cu_4{}^sBu_2(CN)_3^-$, thus indicating the occurrence of reaction but not incorporation of OH⁻ into the cuprate aggregates. It was therefore concluded that the hydrolysis is taking place inside the ion trap. This phenomenon is systematically investigated in Section 4.2.2.

Experiments on THF solutions of $Li_nCuR_n(CN)$, $n = 0.5$ and 0.8 revealed that members of the $Li_{n-1}Cu_nR_n(CN)_n^-$ series and the hydroxyl-containing species $Li_{n-1}Cu_nR_{n-1}(OH)(CN)_n^-$ (R = ^{n}Bu and ^{s}Bu) remain virtually unaffected. In contrast, the cyanide-free $Li_{n-1}Cu_nR_{2n}^-$ anions are reduced in signal intensity with decreasing RLi/CuCN ratios, but in some cases remain visible for $n = 0.8$ and 0.5. Complexes showing the opposite behavior and increasing in relative abundance are $Cu_2R_2(CN)^-$ (observed for R = Me, Et, ^{n}Bu, ^{s}Bu, ^{t}Bu), $LiCu_3R_3(CN)_2^-$ (R = ^{n}Bu and ^{s}Bu), and $Li_2Cu_5R_5(CN)_3^-$ (R = ^{n}Bu and ^{s}Bu). These species have in common R/CN ratios > 1. In no case, complexes with R/CN ratios < 1 are detected. As detailed in Section 6.2.2, samples of a nominal composition of $Li_{0.8}CuR_{0.8}(CN)$ contain LiCuR(CN) in the solution phase because the excess CuCN does not dissolve. In line with this assessment, solutions of $Li_mCuR_m(CN)$ in THF, $m = 0.5$, 0.8, and 1.0, yield very similar ESI mass spectra.

$LiCuR_2 \cdot LiCN$ Solutions in Et$_2$O, MeTHF, CPME, and MTBE. The negative ion mode ESI mass spectra of solutions of $LiCuR_2 \cdot LiCN$ in Et$_2$O are, just as in the case of THF solutions, dominated by complexes of the homologous series $Li_{n-1}Cu_nR_{2n}^-$, $n = 1 - 3$ (Figure 4.1.1.6). Interestingly, however, the distributions obtained for the Et$_2$O solutions are systematically shifted to higher aggregation states compared to their THF counterparts (Table 4.1.1.4). Also note that the absolute ESI signal intensities are considerably lower in the case of the Et$_2$O solutions. For R = ^{t}Bu, the effect of other ethereal solvents was also probed, and $Li_{n-1}Cu_nR_{2n}^-$ anions, $n = 1$ and 2, were again found. While the fraction of the dimeric complex $LiCu_2R_4^-$ is relatively small for MeTHF (2-methyltetrahydrofuran) solutions, this ion apparently prevails in CPME (cyclopentyl methyl ether) and MTBE (methyl *tert*-butyl ether).

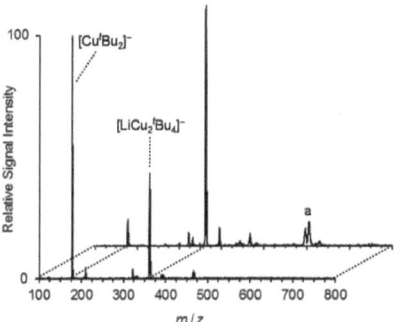

Figure 4.1.1.6. Negative-ion mode ESI mass spectra of 25 mM solutions of LiCu^{t}Bu$_2$·LiCN in THF (front) and Et$_2$O (back). The ion a (m/z 605/607/609) corresponds to Li$_3$Cu$_4$^{t}Bu$_4$(CN)$_4^-$, which presumably results from partial hydrolysis.

Table 4.1.1.4. Organocuprate anions observed upon ESI of Et$_2$O solutions of LiCuR$_2$·LiCN in high [++], medium [+], and low/negligible [–] relative abundance. For comparison, the relative abundances measured for the analogous solutions in THF are given in brackets.

	n	R = Me	Et	^{n}Bu	^{s}Bu	^{t}Bu	Ph
Li$_{n-1}$Cu$_n$R$_{2n}^-$	1	– (–)	– (–)	– (+)	– (++)	+ (++)	+ (++)
	2	++ (++)	– (–)	– (–)	– (–)	++ (+)	– (–)
	3	++ (++)	++ (++)	++ (++)	++ (++)	– (–)	++ (++)

Li$_{0.8}$CuR$_{0.8}$(CN) Solutions in Et$_2$O, MeTHF, CPME, and MTBE. Negative ion mode ESI of solutions of Li$_{0.8}$CuR$_{0.8}$(CN) in Et$_2$O affords a multitude of different organocuprate anions (Figure 4.1.1.7), most of which are already known from analysis of the corresponding THF solutions.

4   Results and Discussion

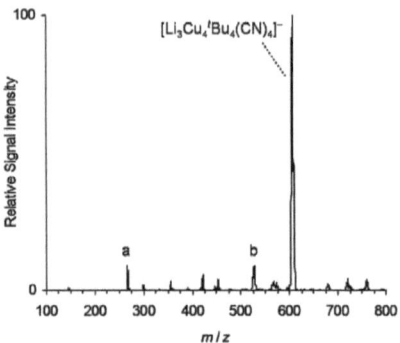

Figure 4.1.1.7. Negative-ion mode ESI mass spectrum of a 25 mM solution of $Li_{0.8}Cu^tBu_{0.8}(CN)$ in Et_2O, a = $Cu_2{}^tBu_2(CN)^-$, b = $Li_3Cu_4{}^tBu_2(OH)_2(CN)_4{}^-$.

A first set of ions belongs to the homologous series $Li_{n-1}Cu_nR_n(CN)_n{}^-$. Ions of this series or the corresponding partially hydrolyzed species $Li_{n-1}Cu_nR_{n-1}(OH)(CN)_n{}^-$ are observed in all cases (Table 4.1.1.5, the partial hydrolysis occurs in ion-molecule reactions with background water present in the ion trap, see Section 4.2.2.). Compared to the situation in THF, the change to Et_2O as solvent does not seem to result in a clear shift in the aggregation states. A second set of prominent ions comprises $Li_{n-1}Cu_nR_{2n}{}^-$ complexes, which do not show the expected stoichiometry but instead are typical of $LiCuR_2 \cdot LiCN$ solutions (see above). The remaining anions observed for solutions of $Li_{0.8}CuR_{0.8}(CN)$ in Et_2O exhibit intermediate stoichiometries and are of limited abundance only. For $Li_{0.8}Cu^tBu_{0.8}(CN)$ in MeTHF, CPME, and MTBE, $Li_{n-1}Cu_nR_n(CN)_n{}^-$ and $Li_{n-1}Cu_nR_{2n}{}^-$ complexes are also observed as the predominating anions.

4   Results and Discussion

Table 4.1.1.5. Organocuprate anions observed upon ESI of Et$_2$O solutions of Li$_{0.8}$CuR$_{0.8}$(CN) in high [++], medium [+], and low/negligible [−] relative abundance. For comparison, the relative abundances measured for the analogous solutions in THF are given in brackets.

	n	R = Me	Et	^{n}Bu	^{s}Bu	^{t}Bu	Ph
Li$_{n-1}$Cu$_n$R$_n$(CN)$_n^-$	2	++ (++)	+ (++)	+ (+)	− (+)	− (−)	++ (++)
	3	− (+)	− (+)	− (+)	− (−)	− (+)	++ (+)
	4	+ (++)	+ (+)	− (−)	− (−)	++ (+)	− (−)
	5	− (+)	− (−)	− (−)	− (−)	− (−)	− (−)
Li$_{n-1}$Cu$_n$R$_{n-1}$(OH)(CN)$_n^-$	3	− (−)	− (−)	− (+)	− (+)	− (−)	− (−)
	4	− (−)	+ (−)	− (+)	++ (++)	− (−)	− (−)
Li$_{n-1}$Cu$_n$R$_{2n}^-$	1	− (−)	− (−)	− (−)	(+)	(++)	++ (+)
	2	− (−)	− (−)	− (−)	− (−)	(+)	− (−)
	3	++ (+)	++ (+)	++ (++)	+ (++)	− (−)	++ (−)
Cu$_2$R$_2$(CN)$^-$		− (−)	− (+)	− (−)	− (−)	− (−)	− (−)
Cu$_2$R$_3^-$		− (−)	− (−)	+ (−)	− (−)	− (−)	− (−)
LiCu$_4$R$_6^-$		− (−)	− (−)	− (−)	− (−)	− (−)	+ (−)

4.1.2. Positive-Ion Mode ESI Mass Spectrometry.

Cyanocuprate Solutions in THF. Positive-ion mode ESI of solutions of LiCuR$_2$·LiCN in THF mainly produces Li$_2$(CN)(THF)$_n^+$, n = 2 and 3, as well as smaller amounts of Li(THF)$_n^+$, n = 2 and 3, and Li$_3$(CN)$_2$(THF)$_2^+$ (see Figure 4.1.2.1 for the case of LiCuMe$_2$·LiCN).

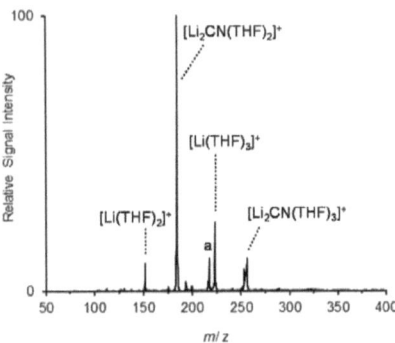

Figure 4.1.2.1. Positive-ion mode ESI mass spectrum of a 25 mM solution of LiCuMe$_2$·LiCN in THF, a = Li$_3$(CN)$_2$(THF)$_2^+$.

In stark contrast, solutions of Li$_m$CuR$_m$(CN), $m \leq 1$, do not yield any Li$_2$(CN)(THF)$_n^+$ (see Figure 4.1.2.2 for the case of Li$_{0.8}$CuMe$_{0.8}$(CN)). Instead, their mass spectra are dominated by Li(THF)$_n^+$, n = 2 and 3, and also show Li$_2$CuR(CN)(THF)$_2^+$ ions in small signal intensities.

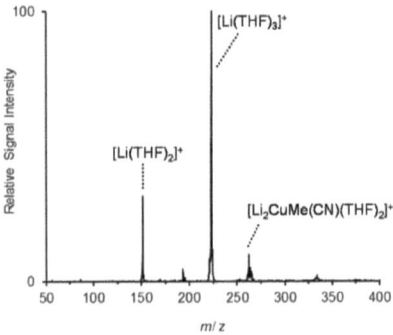

Figure 4.1.2.2. Positive-ion mode ESI mass spectrum of 25 mM Li$_{0.8}$CuMe$_{0.8}$(CN) in THF.

4   Results and Discussion

Cyanocuprate Solutions in Et$_2$O, MeTHF, CPME, and MTBE. Turning to the positive-ion mode ESI mass spectra of solutions of LiCuR$_2$·LiCN in Et$_2$O, Li(Et$_2$O)$_2^+$ and Li$_2$(CN)(Et$_2$O)$_2^+$ are observed as main species (Figure 4.1.2.3). With the exception of R = Et, Li$_2$(CN)(Et$_2$O)$_2^+$ predominates in all cases, thus resembling the situation in THF. For the latter solvent, however, the complexes Li(THF)$_3^+$ and Li$_2$(CN)(THF)$_3^+$ were also detected, whereas ions with n = 3 solvent molecules attached are largely missing in the mass spectra recorded for Et$_2$O solutions. For solutions of LiCuPh$_2$·LiCN in Et$_2$O, incorporation of ^{n}Bu$_2$O in the ions Li$_2$(CN)(Et$_2$O)$_{2-n}$(^{n}Bu$_2$O)$_n^+$, n = 1 and 2 is detected, although the fraction of ^{n}Bu$_2$O in solution (stemming from the preparation of the reagent, see Section 6.2.1) is lower than that of Et$_2$O by a factor of 60. While none of these cations contains a Cu center, small amounts of Li$_2$CuR$_2$(Et$_2$O)$_2^+$ are found for R = Me. The ESI mass spectra measured for solutions of LiCu^{t}Bu$_2$·LiCN in MeTHF, CPME, and MTBE are similar to those obtained for Et$_2$O and THF solutions in that Li(solv)$_n^+$ and Li$_2$(CN)(solv)$_n^+$ are the predominant cations observed.

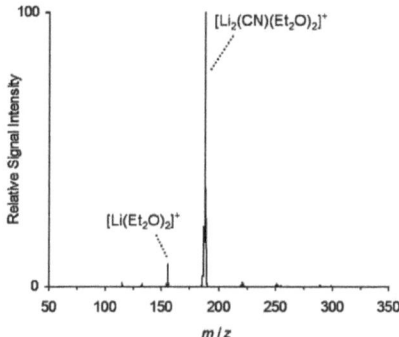

Figure 4.1.2.3. Positive-ion mode ESI mass spectrum of a 25 mM solution of LiCu^{t}Bu$_2$·LiCN in Et$_2$O.

The positive ion mode ESI mass spectra obtained for solutions of Li$_{0.8}$CuR$_{0.8}$(CN) in Et$_2$O in all cases show Li(Et$_2$O)$_2^+$ as the main peak (Figure 4.1.2.4). Very similarly, the corresponding THF solutions also afforded solvated Li$^+$ ions as the predominant species. Less abundant cations observed for the Et$_2$O solutions are Li$_2$(CN)(Et$_2$O)$_2^+$ and Li$_2$CuR(CN)(Et$_2$O)$_2^+$. Solutions of Li$_{0.8}$Cu^{t}Bu$_{0.8}$(CN) in MeTHF also yield Li(solv)$_n^+$ ions as main species (n = 2 and 3), whereas Li$_2$(CN)(solv)$_2^+$ prevails for CPME and MTBE.

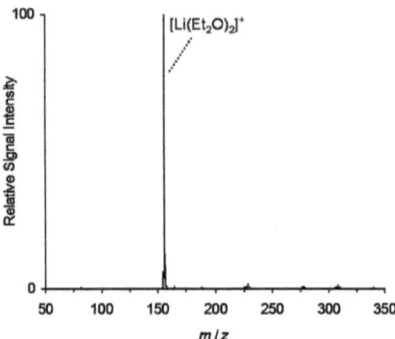

Figure 4.1.2.4. Positive-ion mode ESI mass spectrum of a 25 mM solution of $Li_{0.8}Cu^tBu_{0.8}(CN)$ in Et_2O.

4.1.3. Electrical Conductivity Measurements

LiCuR$_2$·LiCN Solutions. Solutions of LiCuR$_2$·LiCN in THF display significant electrical conductivities (Table 4.1.3.1). The determined molar conductivities show a clear dependence on the nature of the R substituent: Λ(LiCu^{t}Bu$_2$·LiCN) > Λ(LiCu^{n}Bu$_2$·LiCN) > Λ(LiCuPh$_2$·LiCN). The conductivities of LiCuR$_2$·LiCN in Et$_2$O are much smaller, but exhibit a very similar trend.

Table 4.1.3.1. Molar electrical conductivities determined for solutions of LiCuR$_2$·LiCN(c = 91±4 mM) and LiCuR(CN) (c = 97±2 mM) in THF and Et$_2$O at 258 K (activity coefficients are neglected).

R	Molar conductivity Λ(LiCuR$_2$·LiCN)/(S cm^2 mol^{-1})		Molar conductivity Λ(LiCuR(CN))/(S cm^2 mol^{-1})	
	THF	Et$_2$O	THF	Et$_2$O
^{n}Bu	13 ± 1	1.00 ± 0.02	0.3 ± 0.1	0.008 ± 0.005
^{t}Bu	19 ± 1	6.2 ± 0.1	4.0 ± 0.5	0.7 ± 0.1
Ph	8.2 ± 0.3	0.20 ± 0.05	0.42 ± 0.04	0.20 ± 0.05

For LiCuPh$_2$·LiCN in THF, the concentration and temperature dependence was also investigated. At lower concentrations, the molar conductivity increases (Figure 4.1.3.1). A rise in temperature also increases the conductivity (Figure 4.1.3.2). This behavior is expected, because higher temperatures lower the viscosity of the solvent, thus resulting in enhanced ion mobilities. Taking into account Walden's rule, which assumes that the product of the molar conductivity Λ of a given electrolyte and the viscosity of the solvent η is constant,[83] a fit of the measured conductivities on the basis of the known temperature dependence of η(THF) was attempted.[72] The result is reasonably good (Figure 4.1.3.2), suggesting that the observed temperature dependence of the molar conductivity indeed can be rationalized by the change in the viscosity of the solvent.

4   Results and Discussion

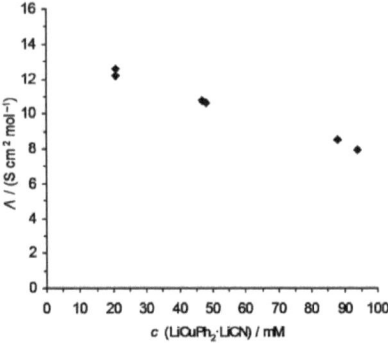

Figure 4.1.3.1. Concentration dependence of the molar conductivity of LiCuPh$_2$·LiCN in THF at T = 258 K.

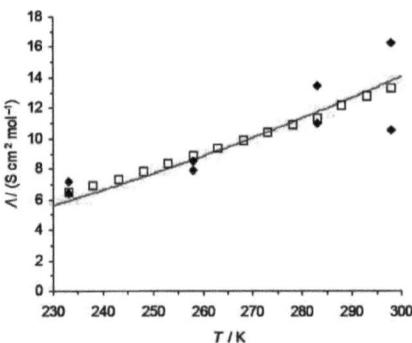

Figure 4.1.3.2. Molar electrical conductivity of a solution of LiCuPh$_2$·LiCN in THF (c = 98 mM, activity coefficients are neglected) as a function of temperature. The open symbols represent data points collected during a single conductivity measurement, in which the temperature was raised continuously from 233 K to 298 K in 5 K increments. The filled symbols represent data points collected independently for different samples at fixed temperatures. The line corresponds to a fit that only takes into account the effect of the temperature dependence of the solvent viscosity.

LiCuR(CN) Solutions. The molar conductivities of solutions of LiCuR(CN) in THF are significantly smaller than those of the corresponding LiCuR$_2$·LiCN solutions (Table 4.1.3.1). Again, a strong dependence on the nature of the R substituent is noticeable: Λ(LiCu^{t}Bu(CN)) $\gg$ Λ(LiCuPh(CN)) > Λ(LiCu^{n}Bu(CN)). The conductivities in Et$_2$O are even lower (also lower than the conductivities measured for Et$_2$O solutions of LiCuR$_2$·LiCN), but show a similar trend. For LiCuPh(CN) in THF, the temperature dependence is reproduced by a simple fit that only considers the effect of the changed solvent viscosity, like in the case of LiCuPh$_2$·LiCN.

4.1.4. General Trends

The results show that ESI mass spectrometry permits the detection of a wide range of non-stabilized organocuprate anions. The successful observation of these highly air- and moisture-sensitive species requires careful sample handling to exclude oxidation and hydrolysis reactions. Possibly, these problems also prevented Lipshutz and coworkers from the observation of intact methyl- and *n*-butylcuprate anions.[37b]

In the absence of oxidation and hydrolysis reactions, the detected organocuprates show a clear dependence on the stoichiometry of the applied reagents. For sample solutions containing CuCN/2 RLi, only cyanide-free anions of the type $Li_{n-1}Cu_nR_{2n}^-$ are observed. This finding completely agrees with the current consensus that diorganocuprates do not form higher-order complexes to a measurable extent (see Section 2.2).[19-21] Sample solutions prepared from CuCN/RLi display a greater manifold of cuprate anions, with $Li_{n-1}Cu_nR_n(CN)_n^-$ complexes being most prominent. The stoichiometry of these species reflects the nominal overall composition of the solution. Furthermore, the incorporation of cyanide in these complexes is in line with the well-known coordination of CN^- to Cu centers in LiCuR(CN) reagents.[21] This accordance with results from the literature gives confidence that the organocuprate anions observed by ESI mass spectrometry indeed are closely related to the solution chemistry of these species.

For LiCuR$_2$·LiCN solutions, the predominance of cyanide-free $Li_{n-1}Cu_nR_{2n}^-$ anions is exactly mirrored by the prevalence of the cyanide-containing $Li_2(CN)(solv)_n^+$ cations. Just the opposite behavior is found for LiCuR(CN) solutions. Here, the cyanide is preferentially incorporated into the anions, whereas simple $Li(solv)_n^+$ complexes predominate for the cations. ESI mass spectrometry thus affords a consistent picture of the partitioning of CN^- ions in cyanocuprates. Note that the present findings seem to disagree with the conclusion of Gschwind, Boche, and coworkers, that the most important solvent separated ion pair in solutions of LiCuR$_2$·LiCN in THF corresponds to $Li(THF)_n^+/CuR_2^-$ (**3**, Scheme 2.2.1).[20,29a,b] In contrast, the observed $Li_2(CN)(THF)_n^+$ cations are closely related to the cationic component of the contact ion pair **2**.[31-34] Possibly, temperature effects may again complicate the situation and make a direct comparison with the conclusions of Gschwind, Boche, and coworkers difficult (see above).

4   Results and Discussion

4.1.5. Equilibria Operative

The detection of $Li_{n-1}Cu_nR_{2n}^-$ anions in $LiCuR_2 \cdot LiCN$ solutions is rationalized by the operation of association/dissociation equilibria, as depicted in Scheme 4.1.5.1. The proposed scenario essentially corresponds to the equilibrium already suggested by Gschwind and collaborators,[20a,29,30] but in addition also accounts for the formation of higher ionic aggregates. While ESI mass spectrometry cannot detect the neutral homodimers $Li_2Cu_2R_4$, the presence of these species in ethereal solutions of $LiCuR_2 \cdot LiCN$ has been proven by NMR spectroscopy, X-ray scattering, and ebullioscopic methods.[20a,29,30,84]

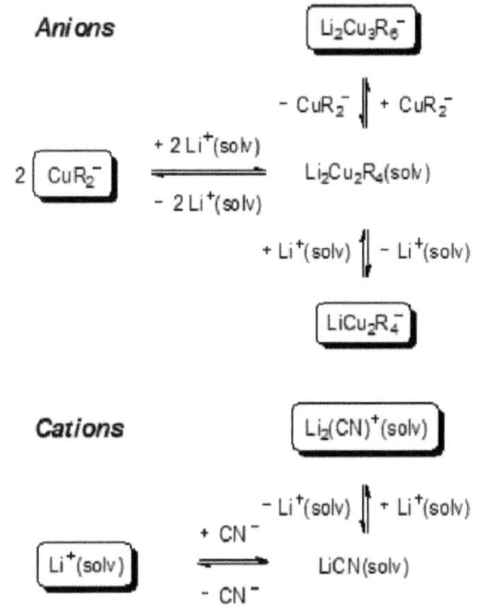

Anions

$$2\,CuR_2^- \; \underset{-2\,Li^+(solv)}{\overset{+2\,Li^+(solv)}{\rightleftharpoons}} \; Li_2Cu_2R_4(solv)$$

with $Li_2Cu_3R_6^-$ formed via $-CuR_2^- / +CuR_2^-$ from $Li_2Cu_2R_4$, and $LiCu_2R_4^-$ formed via $+Li^+(solv) / -Li^+(solv)$.

Cations

$$Li^+(solv) \; \underset{-CN^-}{\overset{+CN^-}{\rightleftharpoons}} \; LiCN(solv)$$

with $Li_2(CN)^+(solv)$ formed via $-Li^+(solv) / +Li^+(solv)$.

Scheme 4.1.5.1. Association/dissociation equilibria proposed to be operative in ethereal solutions of $LiCuR_2 \cdot LiCN$. The highlighted species have been observed by ESI mass spectrometry.

The inferred composition of the $Li_2Cu_3R_6^-$ complexes from $Li_2Cu_2R_4$ and CuR_2^- subunits is also of interest with respect to the observed fast equilibration of $LiCuMe_2 \cdot LiCN$ and $LiCuR_2 \cdot LiCN$ reagents.[85] One obvious exchange process supposedly corresponds to the recombination of subunits with different organyl groups. However, such recombination processes only exchange even numbers of Me for R groups and thus cannot explain the more extensive equilibration found in the experiments (Table 4.1.1.2). Most likely, the exchange

4 Results and Discussion

processes therefore also involve the organyl groups within the $Li_2Cu_2R_4$ and CuR_2^- subunits. As detailed above, the resulting distribution of R and Me groups in the mixed cuprates is asymmetric. The observation of R-enriched anions is ascribed to their reduced affinity toward ion pairing compared to their methyl-rich analogs; this trend parallels the decreased tendency of cuprates with large R groups to form higher aggregates (see above). The notable exception found in the case of R = Ph, for which the $Li_2Cu_3Me_{6-x}Ph_x^-$ ions are depleted in Ph, probably results from the stabilization of the phenyl-enriched contact ion pairs by π-binding of the Ph groups with Li^+ cations. Such π-binding interactions have been reported for related lithium arylcuprates.[86]

For LiCuR(CN) solutions, the situation is less clear because the aggregation state of the neutral, undissociated component [LiCuR(CN)] is not known precisely. Nonetheless, the experimental findings point to the operation of association/dissociation equilibria, such as Eq (4.1.5.1).

$$n\,[LiCuR(CN)] \rightleftharpoons Li^+(solv) + Li_{n-1}Cu_nR_n(CN)_n^- \qquad (4.1.5.1.)$$

Compared to the cyanide-free $Li_{n-1}Cu_nR_{2n}^-$ anions, the $Li_{n-1}Cu_nR_n(CN)_n^-$ complexes show significantly higher aggregation states (Table 4.1.1.3). This deviating behavior is assigned to the presence of CN^- ions, which can more easily adopt bridging binding sites. Reported crystal structures of monoorganocuprates indeed show that CN^- ions connect Cu centers and Li^+ cations.[21] The incorporation of CN^- in the cuprate anions also explains why simple, cyanide-free $Li^+(solv)$ cations prevail for LiCuR(CN), in contrast to the case of $LiCuR_2 \cdot LiCN$. Association of Li^+ with neutral [LiCuR(CN)] rationalizes the formation of $Li_2CuR(CN)^+(solv)$ cations, Eq (4.1.5.2).

$$Li^+(solv) + [LiCuR(CN)] \rightleftharpoons Li_2CuR(CN)^+(solv) \qquad (4.1.5.2.)$$

Fundamental structural differences between $Li_{n-1}Cu_nR_{2n}^-$ and $Li_{n-1}Cu_nR_n(CN)_n^-$ complexes are also evident from their coexistence in solution. The low abundance of mixing products $Li_{n-1}Cu_nR_{n+x}(CN)_{n-x}^-$ is particularly striking in light of the fast exchange observed between $LiCuMe_2 \cdot LiCN$ and $LiCuR_2 \cdot LiCN$ reagents. This finding strongly implies that the organyl and cyanide substituents in the $Li_{n-1}Cu_nR_n(CN)_n^-$ anions adopt non-equivalent binding sites. The presence of $Li_{n-1}Cu_nR_{2n}^-$ complexes in solutions of LiCuR(CN) can possibly be accounted for by Schlenk-type equilibria according to Eq (4.1.5.3). The absence of

$Li_{n-1}Cu_n(CN)_{2n}^-$ anions in the recorded ESI mass spectra might result from their higher tendency to form contact ion pairs with Li^+ relative to their $Li_{n-1}Cu_nR_{2n}^-$ counterparts.

$$2\ LiCuR(CN) \rightleftharpoons LiCuR_2 + LiCu(CN)_2 \qquad (4.1.5.3.)$$

This increased amount of ion pairing for cyanide-rich species is also in line with lower electrical conductivities measured for solutions of LiCuR(CN), as compared to $LiCuR_2 \cdot LiCN$. Also note that the higher dissociation tendency of the latter helps to rationalize why already small amounts of this species present in solutions of LiCuR(CN) can result in appreciable concentrations of $Li_{n-1}Cu_nR_{2n}^-$ anions and their detection by ESI mass spectrometry. As discussed previously, the lower dissociation tendencies of the LiCuR(CN) reagents presumably result from the incorporation of the cyanide in the cuprate species. CN^- not only can bridge different Cu centers but, owing to its ambident nature, at the same time also bind to a Li^+ cation with high affinity, thus causing the build-up of larger aggregates.

4.1.6. Effect of the Solvent

The ESI mass-spectrometric experiments show that the transition from THF to Et$_2$O or other ethereal solvents does not lead to the formation of new ionic species in significant quantities. Instead, it results in a shift in the association/dissociation equilibria for the LiCuR$_2$·LiCN reagents. It is first analyzed whether the observed shift reflects the situation in solution or whether it might mirror different behavior during the ESI process. THF and Et$_2$O not only differ in their polarity but also in their boiling point (Table 4.1.6.1). While the higher polarity of THF is likely to favor dissociation in solution, the lower boiling point of Et$_2$O should facilitate desolvation during the ESI process and thus could also explain the higher propensity to association observed for this solvent. However, the results obtained for LiCu^{t}Bu$_2$·LiCN in MeTHF, CPME, and MTBE clearly show that the solvent polarity is the decisive factor (Table 4.1.6.1).

It is obvious that the Lewis-basic ethereal solvents do not interact with both cations and anions in a similar way, but that they bind to the Li$_n$(CN)$_{n-1}^+$ cations, n = 1 and 2, much more strongly than to the cuprate anions. As a consequence, no micro-solvated cuprate anions are detected by ESI mass spectrometry, whereas exclusively Li$_n$(CN)$_{n-1}$(solv)$_x^+$ cations, x = 2 and 3, are found. Note that the number of bound solvent molecules observed for the gaseous ions presumably does not correspond to the first solvation shell in solution but rather reflects the relative interactions energies (too weakly bound molecules will be lost upon energetic collisions during the ESI process). The higher number of Li$^+$-bound THF and MeTHF molecules ($x \geq 2$) correlates very well with the higher macroscopic polarity of these solvents and their effect of shifting the equilibria toward dissociated ions.

Table 4.1.6.1. Properties of ethereal solvents sampled and aggregation tendencies of Li$_{n-1}$Cu$_n$^{t}Bu$_{2n}^-$ anions in these solvents as determined by ESI mass spectrometry of solutions of LiCu^{t}Bu$_2$·LiCN.

Solvent	Relative permittivity ε(298 K)	Boiling point (K)	I(LiCu$_2$^{t}Bu$_4^-$) / I(Cu^{t}Bu$_2^-$)
THF	7.42[a]	338[a]	< 1
MeTHF	6.97[a]	353[b]	< 1
CPME	4.76[b]	379[b]	> 1
Et$_2$O	4.24[a]	308[a]	> 1
MTBE	2.60[b]	328[b]	> 1

[a]Ref.87 [b]Ref.88

4 Results and Discussion

The conductivity data provide independent and unambiguous evidence that THF favors the dissociation of $LiCuR_2 \cdot LiCN$ in comparison to Et_2O. For the LiCuR(CN) reagents, the trend is much weaker. In line with this observation, the ESI mass spectra measured for solutions of LiCuR(CN) in THF on the one hand and Et_2O on the other do not display notable differences.

4.1.7. Effect of the Organyl Substituent

While all of the LiCuR$_2$·LiCN reagents sampled, as well as their LiCuR(CN) counterparts, behave quite similarly, some differences are discernible. As the conductivity measurements clearly show, the ^{t}Bu substituent favors dissociation and the formation of solvent-separated ion pairs more than the ^{n}Bu and Ph groups. In full accordance with this observation, ESI mass spectrometry finds ^{t}Bu to be the only substituent for which the aggregation state of the Li$_{n-1}$Cu$_n$R$_{2n}^-$ anions is limited to $n \leq 2$. The lower aggregation tendency of the ^{t}Bu-containing cuprates is ascribed to the higher steric demands of this substituent, which apparently prevents the association of > 2 Cu^{t}Bu$_2^-$ monomers. The conductivity data moreover suggest that the ^{n}Bu-bearing homoleptic cuprates give somewhat higher fractions of dissociated ions than their Ph-containing analogs. This finding seems to be at odds with the ESI mass-spectrometric results, which point to a slightly higher aggregation tendency for the ^{n}Bu-bearing cuprates. This discrepancy may possibly arise from the different temperatures in both experiments (258 K for the conductivity measurements and approx. 298 K for the ESI mass-spectrometric studies). However, it could also be the case that the deviating behavior observed by ESI mass spectrometry is due to a (though rather small) perturbation of the system caused by the very ESI process.

The recorded molar conductivities Λ also comprise information on the absolute fractions of dissociated ions. For calculating the degree of dissociation α according to Eq (4.1.7.1), the limiting molar conductivity Λ_0 must be known.

$$\alpha = \Lambda / \Lambda_0 \qquad (4.1.7.1.)$$

Although Λ_0 could be derived by the extrapolation of experimental data to $c = 0$, the very steep slope and the increased susceptibility to inevitable hydrolysis reactions at lowest concentrations render such an approach unreliable. For a rough estimation, the limiting molar conductivities of the lithium cuprates are instead approximated by known values of other electrolytes in THF. At 298 K, the limiting molar conductivities of many diverse 1:1 electrolytes in THF all fall into the range of $75 < \Lambda_0(298\text{ K}) < 135$ S cm^2 mol^{-1},[89] which converts into $\Lambda_0(258\text{ K}) = 65 \pm 20$ S cm^2 mol^{-1} on the basis of Walden's rule.[83] If, simplistically, this value is applied to the lithium cuprates, effective degrees of dissociation of $0.09 \leq \alpha(\text{THF}) \leq 0.44$ and $0.002 \leq \alpha(\text{Et}_2\text{O}) \leq 0.14$ are obtained for solutions of LiCuR$_2$·LiCN at concentrations of $c \approx 100$ mM at 258 K.[90] These estimates indicate that even in the more polar THF the lithium cuprates are far from being completely dissociated. This assessment is also consistent with the ESI mass-spectrometric experiments, which show abundant $Li_{n-1}Cu_nR_{2n}^-$ aggregates in all cases examined.

4.1.8. Effect of the Temperature

From the temperature dependence of the ^{1}H,^{6}Li HOESY coupling observed for solutions of LiCu(CH$_2$SiMe$_3$)$_2$ in THF, John et al. concluded that the association/dissociation equilibrium of this reagent is strongly affected by temperature.[29b] According to the authors, the formation of the solvent-separated ion pairs is enthalpically favored but entropically disfavored because the enhanced solvation of the free Li$^+$ cations results in the loss of degrees of freedom.[29b] For the LiCuPh$_2$·LiCN/THF system, the present measurements show only a modest increase of the conductivity as a function of temperature. The observed increase can be fully explained by the effect of the reduced viscosity and thus excludes a pronounced temperature effect on the association/dissociation equilibrium. This result does not directly disagree with the conclusions of John et al., because the cuprate reagents probed in both studies are different.

4.1.9. Comparison of Analytical Methods

Beyond providing insight specific to the lithium cuprate reagents examined, this investigation also permits a comparison of different experimental methods used for the analysis of ionic species in solution. NMR spectroscopy, electrical conductivity measurements, and ESI mass spectrometry all agree that the association/dissociation equilibria of lithium cyanocuprates are largely governed by the nature of the solvent and, in particular, its Li$^+$ affinity. While NMR spectroscopy and conductometry constitute well-established techniques and thus are expected to give the same results, the consistency of the ESI mass-spectrometric findings deserves some further comments. As mentioned in Section 3.1.1, the mass-spectrometrically detectable ions do not originate directly from the sampled solution but from the intermediately formed nanodroplets. The good agreement between the results obtained by ESI mass spectrometry and by conventional analytical methods indicates that the relative position of the association/dissociation equilibria is largely preserved in the nanodroplets. Note that the ESI mass-spectrometric experiments are sensitive to the nature of the solvent even if the observed ions do not contain any solvent molecules, as the example of the Li$_{n-1}$Cu$_n$R$_{2n}^-$ complexes demonstrates. The observed solvent-dependent shift in the aggregation state of these ions rationalizes at the microscopic level what the conductivity measurements find macroscopically. The consistency between the ESI mass-spectrometric and conductometric results is not limited to the effect of the solvent but also extends to the observation of a particularly high dissociation tendency of the LiCu^{t}Bu$_2$·LiCN reagent. In contrast, a comparison of the data for LiCu^{n}Bu$_2$·LiCN and LiCuPh$_2$·LiCN possibly points to some

4 Results and Discussion

smaller deviations between the two methods and thus seems to suggest that their overall agreement is not absolutely perfect. Nevertheless, the present results demonstrate the suitability of ESI mass spectrometry to probe the speciation of cuprate ions in solution and to provide qualitatively correct insight on their association/dissociation behavior. This assessment is in line with the conclusions of several other recent studies that investigated the performance of ESI mass spectrometry.[59,67]

4 Results and Discussion

4.2. Gas-Phase Reactivity of Cyanocuprates

4.2.1. Gas-Phase Fragmentation Reactions

For all ions detected in sufficient signal intensity, the gas-phase fragmentation behavior was studied (Table 4.2.1.1). Because of the considerable amount of data obtained, only fragmentation channels for the most prominent ions are discussed here.

Table 4.2.1.1. Gas-phase fragmentation reactions of mass-selected organocuprate anions occurring with high (++) and medium (+) branching ratios, pathways not detected are denoted by empty spaces. Reactions not observable because of the absence or insufficient abundance of the parent ions are denoted *n.a.*

entry	parent ion	ionic fragment	neutral fragment	R =	Me	Et	^{n}Bu	^{s}Bu	^{t}Bu	Ph
1	CuR_2^-	$HCuR^-$	(R–H)			++	++	++	++	
2a	$LiCu_2R_4^-$	CuR_2^-	$LiCuR_2$		+	n.a.	n.a.	n.a.	++	++
b		$Cu_2R_3^-$	LiR		++	n.a.	n.a.	n.a.		
3a	$Li_2Cu_3R_6^-$	CuR_2^-	$Li_2Cu_2R_4$		++	++	++	++	n.a.	++
b		$Cu_2R_3^-$	Li_2CuR_3		++		+		n.a.	
4a	$LiCu_2R_2(CN)_2^-$	$LiHCu_2R(CN)_2^-$	(R–H)			++	++	++	++	
b		$LiCuR(CN)_2^-$	CuR		++					
c		$Cu_2R_2(CN)^-$	LiCN		+					++
5a	$Li_2Cu_3R_3(CN)_3^-$	$Cu_2R_2(CN)^-$	$Li_2CuR(CN)_2$		++	++	++	n.a.	++	++
b		$LiCu_2R_2(CN)_2^-$	LiCuR(CN)		+	+	+	n.a.	+	
c		$Li_2Cu_2R_2(CN)_3^-$	CuR					n.a.	+	
6	$Li_3Cu_4R_4(CN)_4^-$	$LiCu_2R_2(CN)_2^-$	$Li_2Cu_2R_2(CN)_2$		++	n.a.	++	n.a.	++	n.a.
7a	$Li_2Cu_3R_2(OH)(CN)_3^-$	$Cu_2R(CN)_2^-$	$Li_2CuR(OH)(CN)$		n.a.	n.a.	++	++	n.a.	n.a.
b		$Cu_2R_2(CN)^-$	$Li_2Cu(OH)(CN)_2$		n.a.	n.a.	+	+	n.a.	n.a.
c		$Cu_3R_2(CN)_2^-$	$Li_2(OH)(CN)$		n.a.	n.a.	+	+	n.a.	n.a.
8	$Li_3Cu_4R_3(OH)(CN)_4^-$	$LiCu_2R_2(CN)_2^-$	$Li_2Cu_2R(OH)(CN)_2$		n.a.	n.a.	++	++	n.a.	n.a.

$Li_{n-1}Cu_nR_{2n}^-$ Anions.

Mononuclear cuprates of the $Li_{n-1}Cu_nR_{2n}^-$ homologous series undergo β-hydrogen eliminations if possible, i.e., if a β-H atom is available (Table 4.2.1.1, entry 1, and Figure 4.2.1.1).

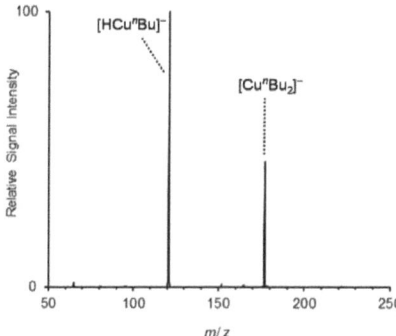

Figure 4.2.1.1. Mass spectrum of mass-selected $Cu^nBu_2^-$ (m/z = 177) and its fragment ions produced upon collision-induced dissociation (V_{exc} = 0.25 V).

This fragmentation channel becomes less important for the higher homologues, which instead preferentially break apart into fragments of reduced nuclearity. So, trimeric $Li_2Cu_3R_6^-$ species lose $Li_2Cu_2R_4$, liberating CuR_2^- as ionic fragment (Table 4.2.1.1, entry 3a, and Figure 4.2.1.2). While the neutral fragments are not directly observable and might undergo further dissociations at higher collision energies, the energetically most favorable pathways should correspond to the formation of intact $Li_2Cu_2R_4$.

4   Results and Discussion

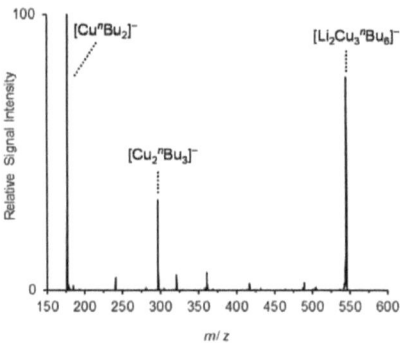

Figure 4.2.1.2. Mass spectrum of mass-selected $Li_2Cu_3{}^nBu_6{}^-$ (m/z = 545) and its fragment ions produced upon collision-induced dissociation (V_{exc} = 0.23 V).

Interestingly, the neutral fragments have exactly the same composition as the contact ion pairs **1** (Scheme 2.2.1), which have been proposed as important constituents of cyanocuprates in ethereal solution.[20,29] The anionic components of the corresponding solvent separated ion pairs **3** in turn are identical with the $CuR_2{}^-$ fragments. This provides additional evidence for the depiction of the observed $Li_2Cu_3R_6{}^-$ complexes as adducts of the contact ion pair **1** with one further $CuR_2{}^-$ anion. In contrast, no indication of polynuclear anions containing the alternatively proposed contact ion pair **2** is found.[31-34]

Decomposition reactions analogous to entries 2 and/or 3 of Table 4.2.1.1 are also observed for the mixed cuprates $Li_{n-1}Me_{2n-x}Cu_nR_x{}^-$. Here, a common feature is the relative depletion of the methyl substituents in the ionic fragments and their corresponding enrichment in the neutral ones (Figure 4.2.1.3).

4 Results and Discussion

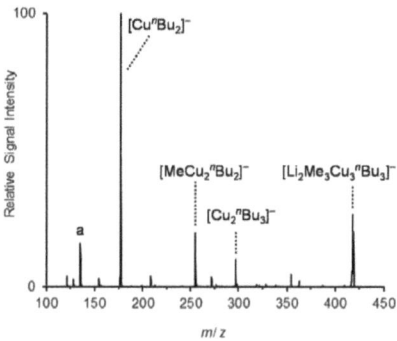

Figure 4.2.1.3. Mass spectrum of mass-selected $Li_2Me_3Cu_3{}^nBu_3{}^-$ (m/z = 419) and its fragment ions produced upon collision-induced dissociation (V_{exc} = 0.16 V), a = $MeCu^nBu^-$.

The neutral fragments lost in most cases contain at least one Cu atom, however, a release of MeLi occurs for $LiCu_2Me_4{}^-$ (Figure 4.2.1.4). The apparent favorability of this fragmentation channel might suggest that lithium methylcuprates could be capable of releasing MeLi in solution as well. Indeed, solutions of $LiCuMe_2$ in a mixture of THF and Et_2O have been shown to yield a positive Gilman test, which is considered indicative of the presence of free RLi species.[91]

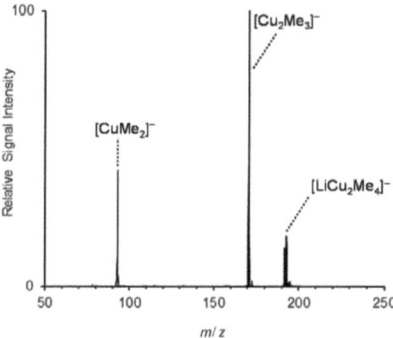

Figure 4.2.1.4. Mass spectrum of mass-selected $LiCu_2Me_4{}^-$ (m/z = 193) and its fragment ions produced upon collision-induced dissociation (V_{exc} = 0.30 V).

4 Results and Discussion

This simple comparison shows how the higher complexity of the polynuclear cuprates opens up additional reaction channels; a similar situation is found for the bimolecular reactivity of lithium organocuprates (Section 4.2.2.).

$Li_{n-1}Cu_nR_n(CN)_n^-$ Anions.

Members of the $Li_{n-1}Cu_nR_n(CN)_n^-$ series show trends similar to those of $Li_{n-1}Cu_nR_{2n}^-$ anions in that β-hydrogen eliminations dominate for the species of lower nuclearity (Table 4.2.1.1, entry 4a, and Figure 4.2.1.5).

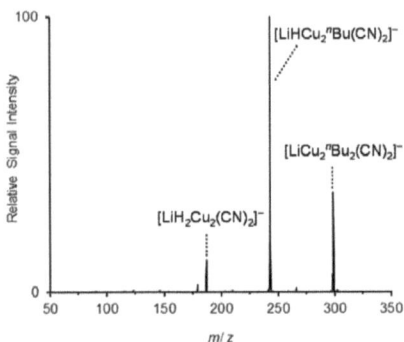

Figure 4.2.1.5. Mass spectrum of mass-selected $LiCu_2{}^nBu_2(CN)_2^-$ (m/z = 299) and its fragment ions produced upon collision-induced dissociation (V_{exc} = 0.35 V).

In the absence of β-H atoms, the dimeric $LiCu_2R_2(CN)_2^-$ complexes afford $LiCuR(CN)_2^-$ (for R = Me) and $Cu_2R_2(CN)^-$ (for R = Me and Ph) as ionic fragments. The $LiCuMe(CN)_2^-$ fragment is special, as it might possibly correspond to a representative of the elusive higher-order cuprates, with a tri-coordinate copper center.

With increasing nuclearity, β-hydrogen eliminations decline and give way to decomposition reactions into smaller aggregates (Table 4.2.1.1, entries 5 and 6, and Figures 4.2.1.6 and 4.2.1.7).

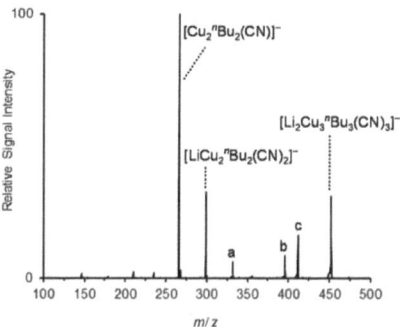

Figure 4.2.1.6. Mass spectrum of mass-selected $Li_2Cu_3{}^nBu_3(CN)_3{}^-$ (m/z = 452) and its fragment ions produced upon collision-induced dissociation (V_{exc} = 0.20 V), a = $Li_2Cu_2{}^nBu_2(CN)_3{}^-$, b = $Li_2HCu_3{}^nBu_2(CN)_3{}^-$, c = $Li_2Cu_3{}^nBu_2(OH)(CN)_3{}^-$. The latter does not correspond to a fragment ion but instead results from in-trap hydrolysis.

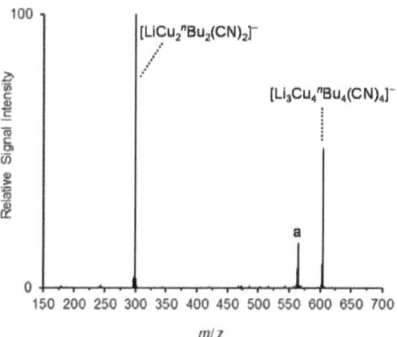

Figure 4.2.1.7. Mass spectrum of mass-selected $Li_3Cu_4{}^nBu_4(CN)_4{}^-$ (m/z = 605) and its fragment ions produced upon collision-induced dissociation (V_{exc} = 0.15 V), a = $Li_3Cu_4{}^nBu_3(OH)(CN)_4{}^-$. The latter does not correspond to a fragment ion but instead results from in-trap hydrolysis.

4 Results and Discussion

In contrast to their homoleptic counterparts, $Li_{n-1}Cu_nR_n(CN)_n^-$ ions yield only very little CuR_2^- or $CuR(CN)^-$ species but instead preferentially form $Cu_2R_2(CN)^-$ and $LiCu_2R_2(CN)_2^-$. At the same time, the energy required to bring about fragmentation decreases for the larger complexes. It is therefore assumed that the $Li_{n-1}Cu_nR_n(CN)_n^-$ complexes do not contain CuR_2^- or $CuR(CN)^-$ as distinct subunits, in line with their reluctance to undergo exchange processes with the $Li_{n-1}Cu_nR_{2n}^-$ anions. As in the case of the latter, analysis of the neutral fragments turns out to be instructive as well. Note again that these species are not observed directly and thus cannot rigorously exclude their further dissociation, which is considered energetically less favorable though. The $Li_2CuR(CN)_2$ fragments produced from $Li_2Cu_3R_3(CN)_3^-$ (Table 4.2.1.1, entry 5a) appear to be related to structure **4**, which Penner-Hahn, Snyder, and coworkers proposed as a minor constituent of mixtures of CuCN/2 LiCl/0.5 MeLi in THF (Scheme 4.2.1.1);[31] compared to **4**, the Cl atom is replaced by a cyanide group. The composition of the $Li_2Cu_2R_2(CN)_2$ fragments produced from $Li_3Cu_4R_4(CN)_4^-$ (Table 4.2.1.1, entry 6) in turn matches that of structure **5**, which forms the predominant motif of solid-state structures of LiCuR(CN) reagents.[21] The present findings suggest that these structures also remain intact in THF solution and form adducts with cuprate anions, thus giving rise to the observed polynuclear $Li_{n-1}Cu_nR_n(CN)_n^-$ complexes.

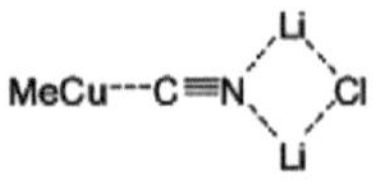

$$MeCu\text{---}C\equiv N\diagup\overset{Li}{\underset{Li}{\diagdown}}\diagdown Cl$$

4

$$RCu\text{---}C\equiv N\diagup\overset{Li}{\underset{Li}{\diagdown}}\diagdown N\equiv C\text{---}CuR$$

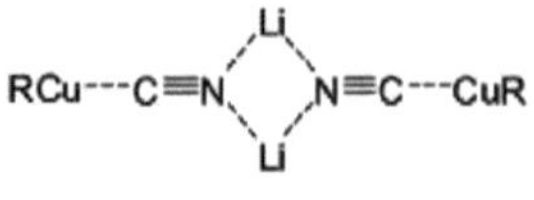

5

Scheme 4.2.1.1. Structures of cyanocuprates reported in the literature.[21,31]

$Li_{n-1}Cu_nR_{n-1}(OH)(CN)_n^-$ Anions.

Just as the parent $Li_{n-1}Cu_nR_n(CN)_n^-$ anions they are derived from, these hydroxyl-containing anions preferentially break apart into clusters of lower nuclearity (Table 4.2.1.1, entries 7 and 8, and Figure 4.2.1.7).

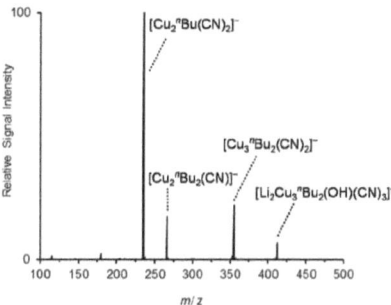

Figure 4.2.1.7. Mass spectrum of mass-selected $Li_2Cu_3{}^nBu_2(OH)(CN)_3{}^-$ ($m/z = 412$) and its fragment ions produced upon collision-induced dissociation ($V_{exc} = 0.20$ V).

Since all neutral fragments lost have the $Li_2(OH)(CN)$ moiety incorporated, an isobaric ^nBu or ^sBu radical loss is theoretically possible. To rule this possibility out and further confirm the assignments, CID experiments on ^{13}C-labeled ions were conducted. These experiments confirmed that $Li_2(OH)(CN)$-containing fragments were lost as neutrals in all cases studied (Table 4.2.1.2).

Table 4.2.1.2. Gas-phase fragmentation reactions of mass-selected $Li_{n-1}Cu_nR_{n-1}(OH)(CN)_n{}^-$ anions, $R = {}^nBu$, ^sBu and $n = 3$ and 4.

Parent ion		Fragment ion		Neutral fragment	
m/z	assignment	m/z	assignment	Δm	assignment
412	$Li_2{}^{63}Cu_3R_2(OH)(CN)_3{}^-$	235	$^{63}Cu_2R(CN)_2{}^-$	177	$Li_2{}^{63}CuR(OH)(CN)$
		266	$^{63}Cu_2R_2(CN)^-$	146	$Li_2{}^{63}Cu(OH)(CN)_2$
		355	$^{63}Cu_3R_2(CN)_2{}^-$	57	$Li_2(OH)(CN)$
415	$Li_2{}^{63}Cu_3R_2(OH)({}^{13}CN)_3{}^-$	237	$^{63}Cu_2R({}^{13}CN)_2{}^-$	178	$Li_2{}^{63}CuR(OH)({}^{13}CN)$
		267	$^{63}Cu_2R_2({}^{13}CN)^-$	148	$Li_2{}^{63}Cu(OH)({}^{13}CN)_2$
		357	$^{63}Cu_3R_2({}^{13}CN)_2{}^-$	58	$Li_2(OH)({}^{13}CN)$
565	$Li_3{}^{63}Cu_4R_3(OH)(CN)_4{}^-$	299	$Li{}^{63}Cu_2R_2(CN)_2{}^-$	266	$Li_2{}^{63}Cu_2R(OH)(CN)_2$
		388	$Li{}^{63}Cu_3R_2(CN)_3{}^-$	177	$Li_2{}^{63}CuR(OH)(CN)$
569	$Li_3{}^{63}Cu_4R_3(OH)({}^{13}CN)_4{}^-$	301	$Li{}^{63}Cu_2R_2({}^{13}CN)_2{}^-$	268	$Li_2{}^{63}Cu_2R(OH)({}^{13}CN)_2$
		391	$Li{}^{63}Cu_3R_2({}^{13}CN)_3{}^-$	178	$Li_2{}^{63}CuR(OH)({}^{13}CN)$

4 Results and Discussion

4.2.2. Gas-Phase Hydrolysis Reactions

Following the observation that selected organocuprate anions can hydrolyze in the ion trap of the instrument, it was decided to characterize the kinetics of this reaction quantitatively. To do so, the ions in question were mass-selected and stored in the ion trap for a given time period Δt, allowing them to react with background water (Scheme 4.2.2.1 for the case of $LiCu_2Me_4^-$). The resulting ionic products, together with remaining parent ions, were then analyzed.

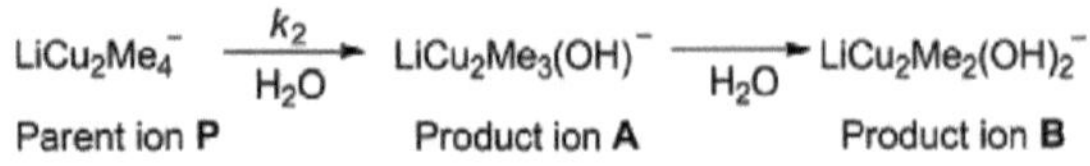

$$LiCu_2Me_4^- \xrightarrow[H_2O]{k_2} LiCu_2Me_3(OH)^- \xrightarrow[H_2O]{} LiCu_2Me_2(OH)_2^-$$

Parent ion **P**   Product ion **A**   Product ion **B**

$-d(P)/dt = k_2[H_2O][P] = k_1[P]$, if $[H_2O]$ = const. $k_2 = k_1/[H_2O]$

define $I_0 = I_P + I_A + I_B$, then $\ln(I_P/I_0) = k_1 t$

Scheme 4.2.2.1. Kinetic model used for determining experimental hydrolysis rate constants.

Given that the concentration of water remains constant, pseudo-first order reaction kinetics can be expected (Scheme 4.2.2.1). Assuming that no ions are lost from the trap and no neutral Cu-containing products are formed, the initial parent ion intensity I_0 should be equal to the sum of intensities of all ions detected (Scheme 4.2.2.1), which allows to calculate the normalized intensity at any point in time Δt. If all the above assumptions hold, plots of the logarithm of the normalized parent ion intensity vs. reaction time should be linear for all ions in question. This was indeed found to be the case (Figure 4.2.2.1 for $LiCu_2Me_4^-$).

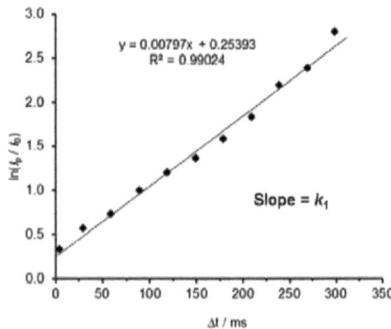

Figure 4.2.2.1. Plot of the logarithm of the normalized ion intensity vs. reaction time for $Li^{63}Cu_2Me_4^-$ (m/z = 193). The non-zero intercept of the fit is due to the fact that hydrolysis occurs already during the isolation of the precursor ion inside the trap.

4 Results and Discussion

The described approach yields pseudo first order rate constants k_1, determined from the slope of the linear fit (Table 4.2.2.1). Several independent measurements were done on different days to ensure reproducibility of data. For reactions that were too slow to be followed, an upper k_1 limit was calculated. So, for the whole set of the nearly inert ions in question, the one that reacted fastest ($Li_2Cu_3Ph_6^-$) was selected and the total product intensity at maximal reaction time was determined. The equation in Scheme 4.2.2.1 was then used to estimate k_1. The respective values for other ions were then stated not to exceed this limit.

To make comparison easier, the slowest reaction, for which the k_1 value could be accurately determined, was chosen as a standard (namely the hydrolysis of $LiCu_2Ph_4^-$). This reaction ($k_1 = 0.089$ s^{-1}) was assigned a *relative* first-order rate constant k_{rel} of unit value, and the rate constants of all other reactions were scaled with respect to it, so that $k_{rel}(\mathbf{A}) = k_1(\mathbf{A})/0.089$ s^{-1}.

Table 4.2.2.1. Gas-phase hydrolysis rate constants of mass-selected organocuprate anions, in relative units[a]. Reactions not observable due to absence of the parent ions, or associated technical difficulties are denoted *n.a.*

Parent ion	n	R = Me	Et	^{n}Bu	^{s}Bu	^{t}Bu	Ph
$Li_{n-1}Cu_nR_{2n}^-$	1	≤ 0.02	*n.a.*	*n.a.*	0.9 ± 0.1	2.1 ± 0.3	≤ 0.02
	2	88 ± 7.0	*n.a.*	*n.a.*	*n.a.*	48 ± 6	**1.0 ± 0.1**
	3	3.5 ± 0.4	8 ± 1	5.3 ± 0.4	17 ± 2.5	*n.a.*	≤ 0.02
$LiCu_2{}^tBu_{4-n}R_n^-$	1	31 ± 3	90 ± 23	*n.a.*	*n.a.*	48 ± 6	15 ± 2
	2	34 ± 5	*n.a.*	*n.a.*	*n.a.*	48 ± 6	4.3 ± 0.7
$Li_{n-1}Cu_nR_n(CN)_n^-$	2	≤ 0.02	*n.a.*	≤ 0.02	≤ 0.02	≤ 0.02	≤ 0.02
	3	9.6 ± 1	*n.a.*	56 ± 17	*n.a.*	16 ± 3	0.6 ± 0.2
	4	*n.a.*	*n.a.*	*n.a.*	*n.a.*	8.5 ± 2	*n.a.*
$LiCu_2R_3(CN)^-$		33 ± 8	*n.a.*	*n.a.*	*n.a.*	19 ± 2	2.7 ± 0.4
$LiCu_2R_3(OH)^-$		*n.a.*	*n.a.*	*n.a.*	*n.a.*	10 ± 1.4	*n.a.*
$Li_2Cu_3R_4(CN)_2^-$		*n.a.*	*n.a.*	*n.a.*	*n.a.*	*n.a.*	19 ± 2
$Li_2CuR(OH)(L^b)_2^+$		39 ± 5	*n.a.*	*n.a.*	*n.a.*	*n.a.*	*n.a.*
$Li_2CuR(CN)(L)_2^+$		9 ± 3	*n.a.*	*n.a.*	*n.a.*	*n.a.*	*n.a.*
$Cu_2R_2(CN)^-$		≤ 0.02	*n.a.*	≤ 0.02	≤ 0.02	≤ 0.02	≤ 0.02

[a] The hydrolysis rate constant of $LiCu_2Ph_4^-$ ($k_1 = 0.089$ s^{-1}) was assigned unit value. [b] L = THF.

4 Results and Discussion

The concentration of water was estimated at $(8 \pm 2) \cdot 10^{10}$ molecule·cm^{-3} (see Section 6.3 of the Appendix), which translates into $p(H_2O) = (3 \pm 0.8) \cdot 10^{-6}$ mbar at 298 K (cf. estimated pressure of helium of ≈ 3 mbar).

On the basis of the data collected, some instructive conclusions can be drawn. Firstly, the exact reaction pathway depends on the aggregation state and composition of the ion in question (Scheme 4.2.2.2), but not on the nature of R group. Secondly, the presence of a Li center greatly enhances the reaction rate, as can be seen from the comparison of CuR_2^- with $LiCu_2R_4^-$ and $Li_2Cu_3R_6^-$ systems (Table 4.2.2.1).

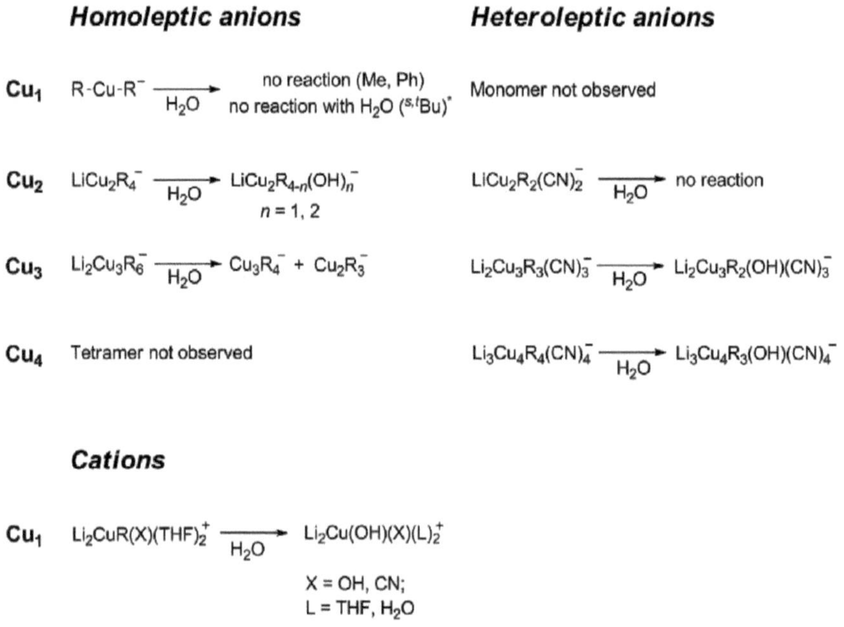

Homoleptic anions

Cu_1 $R\text{-}Cu\text{-}R^-$ $\xrightarrow{H_2O}$ no reaction (Me, Ph)
                                    no reaction with H_2O ($^{s,t}Bu$)*

Cu_2 $LiCu_2R_4^-$ $\xrightarrow{H_2O}$ $LiCu_2R_{4-n}(OH)_n^-$
                                               $n = 1, 2$

Cu_3 $Li_2Cu_3R_6^-$ $\xrightarrow{H_2O}$ $Cu_3R_4^-$ + $Cu_2R_3^-$

Cu_4 Tetramer not observed

Heteroleptic anions

Cu_1 Monomer not observed

Cu_2 $LiCu_2R_2(CN)_2^-$ $\xrightarrow{H_2O}$ no reaction

Cu_3 $Li_2Cu_3R_3(CN)_3^-$ $\xrightarrow{H_2O}$ $Li_2Cu_3R_2(OH)(CN)_3^-$

Cu_4 $Li_3Cu_4R_4(CN)_4^-$ $\xrightarrow{H_2O}$ $Li_3Cu_4R_3(OH)(CN)_4^-$

Cations

Cu_1 $Li_2CuR(X)(THF)_2^+$ $\xrightarrow{H_2O}$ $Li_2Cu(OH)(X)(L)_2^+$

                                                     X = OH, CN;
                                                     L = THF, H_2O

Scheme 4.2.2.2. Summary of hydrolysis pathways followed by the organocuprate ions investigated.

The effect of substituting R for CN depends wholly on the aggregation state of the ion in question. So, $LiCu_2R_4^-$ anions are more reactive than $LiCu_2R_3(CN)^-$, which, in turn, react faster than the completely inert $LiCu_2R_2(CN)_2^-$ (Table 4.2.2.1). On the contrary,

---

* For these anions, a reaction with background O_2 seems to be taking place.

4 Results and Discussion

$Li_2Cu_3R_3(CN)_3^-$ ions are hydrolyzed faster than their homoleptic $Li_2Cu_3R_6^-$ counterparts (Table 4.2.2.1). Mixed dimeric aggregates $LiCu_2{}^tBu_{4-n}R_n^-$, R = Et, Me, Ph and n = 1, 2 provide further insight into the reaction mechanism (Table 4.2.2.2). For $LiCu_2{}^tBu_3R^-$ systems, the R = Et group is hydrolyzed in preference to ^tBu, in the case of R = Me the two rates are comparable, whereas for R = Ph the tert-butyl group is hydrolyzed in preference. In $LiCu_2{}^tBu_2R_2^-$ systems, however, both Me and Ph are preferentially hydrolyzed. To account for these observations, a mechanistic model was developed (Scheme 4.2.2.3).

Table 4.2.2.2. Gas-phase hydrolysis of mixed ^tBu dimer anions. Major product ions are denoted by [+]. Reactions not observable due to absence of the parent ions, are denoted n.a.

Parent ion	Product ion	R = Me	Et	Ph
$LiCu_2{}^tBu_3R^-$	$LiCu_2{}^tBu_3(OH)^-$	+	+	
	$LiCu_2{}^tBu_2R(OH)^-$	+		+
	$LiCu_2{}^tBu_2(OH)_2^-$	+		
$LiCu_2{}^tBu_2R_2^-$	$LiCu_2{}^tBuR_2(OH)^-$		n.a.	
	$LiCu_2{}^tBu_2R(OH)^-$	+	n.a.	+
	$LiCu_2{}^tBu_2(OH)_2^-$	+	n.a.	

The structures of the parent anions in question depend on their aggregation state. So, it is assumed that the structures of trimeric $Li_2Cu_3R_6^-$ ions are similar to that of $Li_2Cu_3Ph_6^-$, which has been characterized by X-ray crystallography.[92] The corresponding dimeric $LiCu_2R_4^-$ species are thought to have a geometry analogous to that of $LiCu_2Me_4^-$, which has been studied by theoretical calculations[54]. Both proposed structures, together with the hydrolysis mechanism, are represented in Scheme 4.2.2.3.

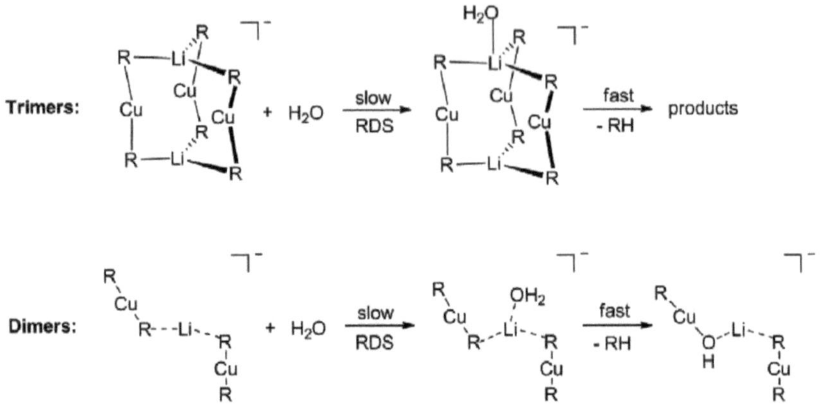

4 Results and Discussion

Scheme 4.2.2.3. Proposed hydrolysis mechanism of dimeric and trimeric anions observed, with the coordination of water being the rate-determining step (RDS).

In the first step, water binds to the coordinatively unsaturated lithium cation. Subsequent fast proton transfer, and, in some cases, fragmentations then lead to the observed products. The fact that in no case studied the anion-water complex was observed implies that it decomposes very fast when formed, i.e., that coordination of water is the rate-determining step (RDS). Another implication of this mechanism is that only the R groups directly coordinating to Li can be hydrolyzed. For example, consider the series $LiCu_2R_4^-$ – $LiCu_2R_3(CN)^-$ – $LiCu_2R_2(CN)_2^-$ (structures given in Scheme 4.2.2.4), with reaction rates decreasing from left to right. This is rationalized by the fact that the first member of the series, $LiCu_2R_4^-$, has two strongly basic R groups next to Li, both of which can be hydrolyzed. The second member, $LiCu_2R_3(CN)^-$, has only one such group, and $LiCu_2R_2(CN)_2^-$ has none (cyano groups being far less basic than R), and is inert towards gas-phase hydrolysis.

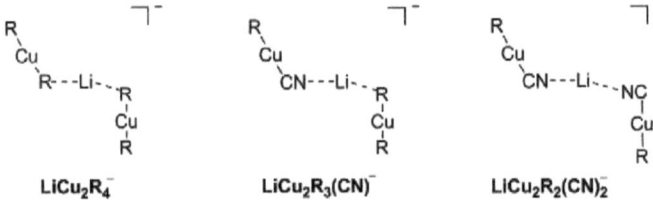

Scheme 4.2.2.4. Suggested structures for selected dimeric organocuprate anions.

For LiCu$_2{}^t$Bu$_{4-n}$R$_n{}^-$ systems, it is argued that the propensity of Me, Et, and Ph groups to coordinate to Li is larger than that of the bulky ^{t}Bu group. Small alkyl groups (Me and Et) coordinate better due to their small size, and, hence, stronger electrostatic interactions. The phenyl group is comparatively large, but it can coordinate to the Li$^+$ center via its π-electron system, and is therefore preferred to the ^{t}Bu group. Due to these preferences, the lithium in LiCu$_2{}^t$Bu$_3$R$^-$ anions should be preferentially coordinated by one ^{t}Bu and one R group, and by two R groups in LiCu$_2{}^t$Bu$_2$R$_2{}^-$. Let us consider the former system. The propensity of an organic group for hydrolysis will be governed by its gas-phase basicity, which decreases in the row Et > Me ≈ ^{t}Bu > Ph.[93] This is in full accord with the experimental results (Table 4.2.2.2). For the case of LiCu$_2{}^t$Bu$_2$R$_2{}^-$ systems, only the R groups can be hydrolyzed. The fact that minor products corresponding to the hydrolysis of ^{t}Bu groups are detected means that species with a ^{t}Bu–Li–R coordination motif are present, or that a rearrangement of the complex takes place in the course of the reaction.

Similarly, this mechanism can explain why trimeric Li$_2$Cu$_3$R$_6{}^-$ anions are hydrolyzed slower than their dimeric counterparts, LiCu$_2$R$_4{}^-$. The Li center in the dimer is more coordinatively unsaturated, and less sterically crowded, than in the trimeric structure (Scheme 4.2.2.3). Hence, coordination of water to the former is both faster and more energetically favorable, which results in faster hydrolysis. Finally, comparing Li$_2$Cu$_3$R$_6{}^-$ with Li$_2$Cu$_3$R$_3$(CN)$_3{}^-$, the decreased number of bulky R groups is believed to reduce the steric hindrance around the three-coordinate Li and thus favors the coordination of water and faster hydrolysis of the latter.

4 Results and Discussion

4.3. Cross-Coupling Reactions

4.3.1. Reactions of Dialkylcuprates with Organyl Halides

Reactions of Dimethylcuprate with Alkyl Halides. Upon addition of 1 equiv of allyl chloride to a solution of LiCuMe$_2$·LiCN (**6**) in THF, the electrical conductivity markedly decreases (Figure 4.3.1.1). At the same time, the ESI signal intensities of the Li$_{n-1}$Cu$_n$Me$_{2n}^-$ ions characteristic of solutions of **6** in THF almost completely vanish.

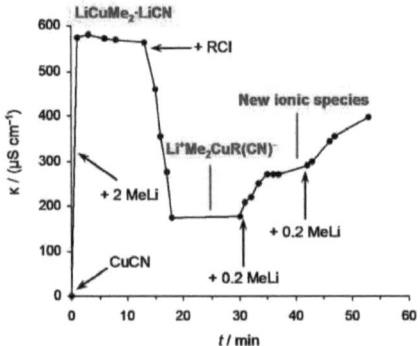

Figure 4.3.1.1. Time profile of the electrical conductivity of a solution of LiCuMe$_2$·LiCN (**6**) in THF (generated by the addition of 2 equiv of MeLi to CuCN) at 202 K upon consecutive treatment with RCl (R = allyl, 1 equiv) and MeLi (2 × 0.2 equiv).

According to Bertz et al.,[44b] this behavior is rationalized by the generation of an Li$^+$Me$_2$CuR(CN)$^-$ intermediate (R = allyl). Due to its relatively low stability,[45c] the Me$_2$CuR(CN)$^-$ anion presumably does not survive the ESI process, thus explaining the inability to detect it by ESI mass spectrometry. If further MeLi is added, the conductivity slowly increases again, indicating the formation of a new ionic species (Figure 3.3.1.1). Similar results are obtained when CuCN/3 MeLi is treated with RCl (Figure 4.3.1.2). In this case, the conductivity first sharply drops, but then slowly recovers as the transient Li$^+$Me$_2$CuR(CN)$^-$ reacts with excess MeLi present in solution to yield the ionic species already known from the previous experiment.

4 Results and Discussion

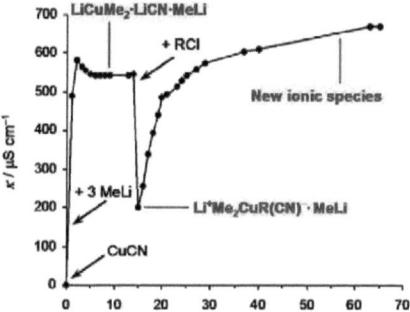

Figure 4.3.1.2. Time profile of the electrical conductivity of a solution of **6** in THF (generated by the addition of 3 equiv of MeLi to CuCN) at 202 K upon treatment with RCl (R = allyl, 1 equiv).

ESI mass spectrometry identifies the newly formed ionic species as the tetraalkylcuprate Me_3CuR^- (Figure 4.3.1.3), which apparently originates from $Li^+Me_2CuR(CN)^-$ via a methide/cyanide exchange (Scheme 4.3.1.1). In addition to mononuclear Me_3CuR^-, the corresponding triple ion, i.e., the Li^+-bound dimer $LiMe_6Cu_2R_2^-$ is also observed. The aggregation equilibrium interrelating mononuclear tetraalkylcuprates and the related dimeric complexes will be discussed in detail in Section 4.3.2.

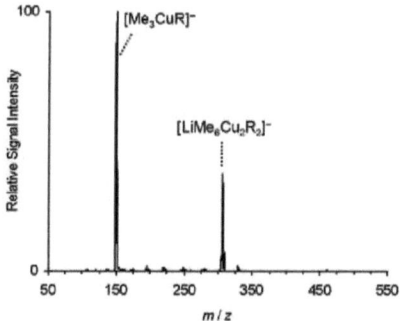

Figure 4.3.1.3. Negative-ion mode ESI mass spectrum of a solution of the products formed in the reaction of $LiCuMe_2 \cdot LiCN$ (**6**) with RCl (R = allyl) in THF.

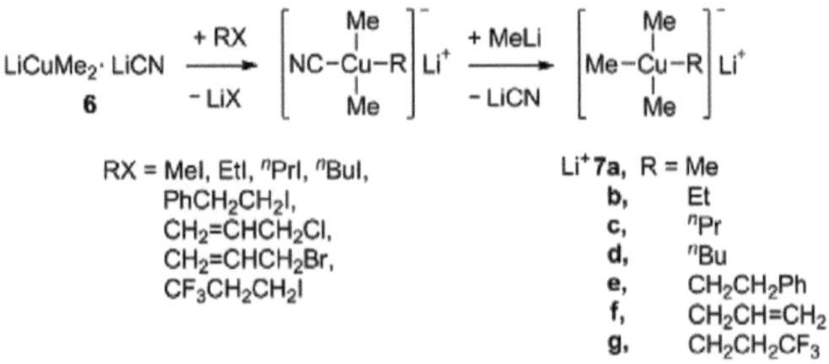

4   Results and Discussion

$$\text{LiCuMe}_2 \cdot \text{LiCN} \quad \xrightarrow[-\text{LiX}]{+\text{RX}} \quad \left[\begin{array}{c} \text{Me} \\ | \\ \text{NC-Cu-R} \\ | \\ \text{Me} \end{array} \right]^{-} \text{Li}^{+} \quad \xrightarrow[-\text{LiCN}]{+\text{MeLi}} \quad \left[\begin{array}{c} \text{Me} \\ | \\ \text{Me-Cu-R} \\ | \\ \text{Me} \end{array} \right]^{-} \text{Li}^{+}$$
6

RX = MeI, EtI, ^{n}PrI, ^{n}BuI,
PhCH$_2$CH$_2$I,
CH$_2$=CHCH$_2$Cl,
CH$_2$=CHCH$_2$Br,
CF$_3$CH$_2$CH$_2$I

Li$^+$**7a**, R = Me
   b, Et
   c, ^{n}Pr
   d, ^{n}Bu
   e, CH$_2$CH$_2$Ph
   f, CH$_2$CH=CH$_2$
   g, CH$_2$CH$_2$CF$_3$

Scheme 4.3.1.1. Formation of lithium tetraalkylcuprates Li$^+$**7a-g** probed by ESI mass spectrometry.

Analogous ESI mass spectrometric experiments demonstrate that **6** not only reacts with allyl chloride, but also with MeI, EtI, ^{n}PrI, ^{n}BuI, PhCH$_2$CH$_2$I, and CH$_2$=CHCH$_2$Br to yield tetraalkylcuprates Me$_3$CuR$^-$ (**7a-f**) and LiMe$_6$Cu$_2$R$_2^-$ (**8a-f**, Scheme 4.3.1.1). Note that **7a**, **7b**, and **7f** are already known from NMR-spectroscopic experiments,[44b,d-f, 45a,c] whereas **7c-e** and **g** have not been reported before. The given assignments are based on observed *m/z* ratios, isotopic patterns, and, for selected systems, isotopic labeling experiments. Additional and unambiguous evidence for the identities of the tetraalkycuprate ions comes from their fragmentation behavior (see Section 4.3.3). Solutions of Li$^+$**7b-g** kept at room temperature are stable for approx. 1h, after which time Cu(I) decomposition products containing CN$^-$ and I$^-$ start to appear. In the case of Li$^+$**7a**, such decomposition products are observed already immediately after sample preparation. The putative Cu(III) species formed upon reaction of **6** with CH$_2$=CHCH$_2$I and BnBr (Bn = benzyl), respectively, prove to be even less stable and completely elude detection by ESI mass spectrometry. In contrast, ^{n}PrCl, BnCl, ^{n}PrBr, (CH$_3$)$_3$CCH$_2$Br, and ^{i}PrI do not react with **6** at all. From these findings, the following trends in reactivity can be derived: (i) Alkyl iodides react faster than the corresponding bromides (compare, e.g., ^{n}PrI and ^{n}PrBr), whereas the chlorides are even less reactive (compare, e.g., BnBr and BnCl). (ii) Primary alkyl halides react faster than secondary ones (compare, e.g., ^{n}PrI and ^{i}PrI). This behavior matches that of typical S$_N$2 processes[94] and thus strongly suggests that the reaction of **6** with alkyl halides follows the same mechanism, in line with previous conclusions.[19a,41a,44b]

A special situation is found for the reaction of **6** with 3,3,3-trifluoropropyl iodide. This reaction affords only small quantities of the expected tetraalkylcuprate **7g**, but mainly gives $Me_{4-n}CuR_n^-$ ions, $R = CF_3CH_2CH_2$ and $n = 2 - 4$ (Figure 4.3.1.4).

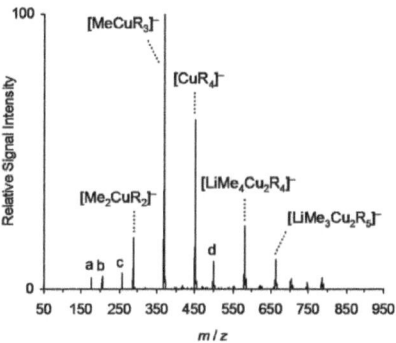

Figure 4.3.1.4. Negative-ion mode ESI mass spectrum of a solution of the products formed in the reaction of $LiCuMe_2 \cdot LiCN$ with RI ($R = CF_3CH_2CH_2$) in THF, a = $MeCuR^-$, b = Me_3CuR^-, c = CuR_2^-, d = $LiMe_5Cu_2R_3^-$.

The formation of these species is explained by the operation of iodine-copper exchange reactions between **6** and RI leading to $LiCu(Me)R \cdot LiCN$ and $LiCuR_2 \cdot LiCN$ reagents (Scheme 4.3.1.2), which can undergo sequential R/CN and R/Me exchanges with the primary $Me_2CuR(CN)^-$ intermediate to yield the observed $Me_{4-n}CuR_n^-$ ions. Support for this rationalization is provided by the observation of $MeCuR^-$ and CuR_2^- (Figure 4.3.1.4). The increased tendency of $CF_3CH_2CH_2I$ to undergo iodine-copper exchange reactions obviously results from the electron-withdrawing effect of the terminal CF_3 group, which helps to stabilize the exchanged cuprates by a better delocalization of the negative charge. Interestingly, iodine-copper exchange also and exclusively occurs for the reaction of **6** with neopentyl iodide (Scheme 4.3.1.2). Apparently, the copper-iodine exchange is less sensitive to steric constraints than an S_N2 reaction and therefore prevails over the latter for the relatively bulky neopentyl system.

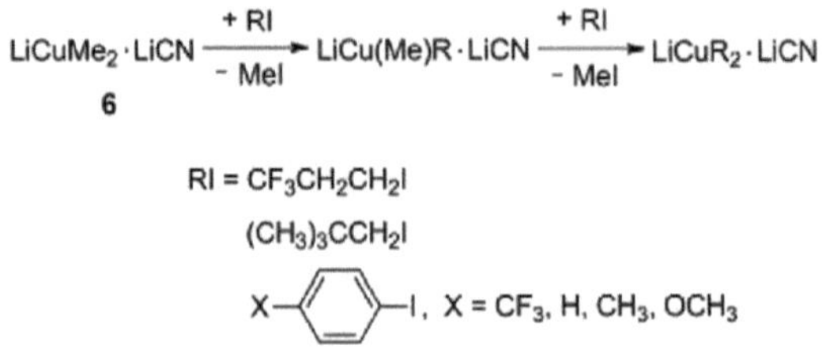

4   Results and Discussion

$$\text{LiCuMe}_2 \cdot \text{LiCN} \xrightarrow[-\text{MeI}]{+\text{RI}} \text{LiCu(Me)R} \cdot \text{LiCN} \xrightarrow[-\text{MeI}]{+\text{RI}} \text{LiCuR}_2 \cdot \text{LiCN}$$
6

RI = $CF_3CH_2CH_2I$
    $(CH_3)_3CCH_2I$
    X—⟨C_6H_4⟩—I, X = CF_3, H, CH_3, OCH_3

Scheme 4.3.1.2. Iodine-copper exchange reactions probed by ESI mass spectrometry.

Reactions of Dimethylcuprate with Aryl Halides. Analysis of mixtures of **6** and aryl iodides RI by negative ion mode ESI mass spectrometry shows the formation of R-bearing cuprates(I) $\text{Li}_{n-1}\text{Cu}_n\text{Me}_{2n-x}\text{R}_x^-$ (Figures 4.3.1.5 and 4.3.1.6). These species originate from sequential iodine-copper exchange reactions (Scheme 4.3.1.2 for $n = 1$), which are well-known to occur upon treatment of dialkylcuprates with aryl halides.[41a,95] The driving force of these processes again is the better stabilization of the negative charge of the cuprate anions by the sp^2-hybridized and, thus, more electron-withdrawing carbon atoms of the aryl groups. The relative stability of the resulting aryl-containing cuprates can be further fine-tuned by changing their electronic properties: Compared to simple phenyl, acceptor-substituted pentafluorophenyl and *p*-trifluoromethylphenyl enhance the stability, whereas donor-substituted *p*-tolyl and, in particular, *p*-anisyl groups reduce it.

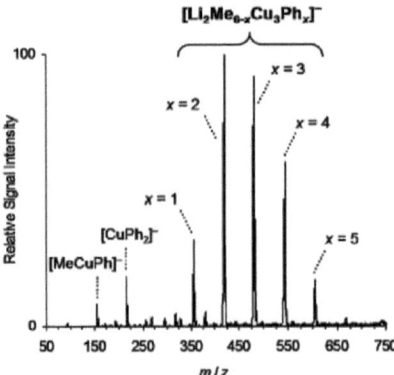

Figure 4.3.1.5. Negative-ion mode ESI mass spectrum of a solution of the products formed in the reaction of LiCuMe$_2$·LiCN (**6**) with PhI in THF.

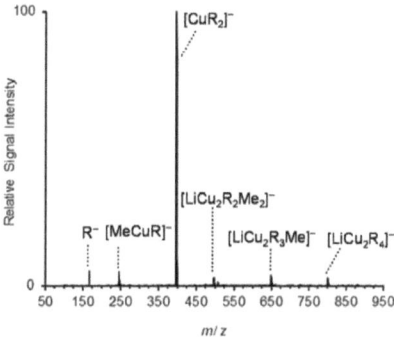

Figure 4.3.1.6. Negative-ion mode ESI mass spectrum of a solution of the products formed in the reaction of LiCuMe$_2$·LiCN (**6**) with RI in THF, R = C$_6$F$_5$.

The halogen-copper exchange reactions of **6** with aryl bromides and chlorides was also investigated. Mixtures of **6** and PhBr give ESI mass spectra essentially identical to those obtained for **6**/PhI. In contrast, **6** does not react with PhCl at room temperature. Similar reactivity orders are known for many other halogen-metal exchange reactions.[96]

Further Reactions of Dialkylcuprates with Organyl Halides. Generation of tetraalkylcuprates anions other than **7** (and the related triple ions **8**) and their detection by ESI mass spectrometry was also attempted. The most obvious way to do so appears to be the reaction of diorganylcuprates LiCuR$_2$·LiCN (R ≠ Me) with alkyl halides R'X, which should afford Li$^+$R$_3$CuR'$^-$ species in analogy to the mechanism depicted in Scheme 4.3.1.1. In no case examined, however, was this approach successful (Table 4.3.1.1).

4 Results and Discussion

Table 4.3.1.1. Reactions of further diorganylcuprates LiCuR$_2$·LiCN with alkyl halides R'X in THF as observed by negative-ion mode ESI mass spectrometry.

Cuprate Reagent	Alkyl Halide	Observed Reaction
LiCuEt$_2$·LiCN	MeI	Decomposition^a
LiCu^{n}Bu$_2$·LiCN	MeI	Iodine-copper exchange
LiCu^{n}Bu$_2$·LiCN	CH$_2$=CHCH$_2$Br	No reaction
LiCu^{t}Bu$_2$·LiCN	MeI	Iodine-copper exchange (slow)
LiCuPh$_2$·LiCN	MeI	Decomposition^a
LiCuPh$_2$·LiCN	^{n}BuI	No reaction
LiCu(Me)^{n}Bu·LiCN	MeI	No reaction
LiCu(Me)^{n}Bu·LiCN	^{n}BuI	No reaction

^{a} Affords cyanide-containing Cu(I) product ions: LiCu$_2$R$_2$(CN)$_2^-$, Li$_2$Cu$_2$R$_2$(CN)$_2$I$^-$ (R = Ph, Et), Li$_2$Cu$_3$Ph$_3$(CN)$_3^-$, Cu$_2$Et$_2$CN$^-$, LiCuEt(CN)I$^-$, LiCu$_2$Et(CN)$_2$I$^-$.

As an alternative method to prepare further tetraalkylcuprates, **6** was treated with ^{n}BuI to generate a Li$^+$Me$_2$Cu^{n}Bu(CN)$^-$ intermediate as described above. Addition of 1 equiv of ^{n}BuLi then yields Me$_2$Cu^{n}Bu$_2^-$ via an ^{n}Bu/CN exchange, though apparently in rather small amounts. An analogous sequential treatment of **6** with ^{n}PrI and ^{n}BuLi affords Me$_2$Cu^{n}Pr(^{n}Bu)$^-$, but in even lower abundance than in the case of its Me$_2$Cu^{n}Bu$_2^-$ counterpart. A more efficient access to Cu(III) species containing three different alkyl substituents was found for the triple ions. Such LiMe$_6$Cu$_2$R(R')$^-$ complexes can be prepared by combination of **6** with 0.5 equiv of RI and 0.5 equiv of R'I, with R/R' = Et, ^{n}Pr, ^{n}Bu, PhCH$_2$CH$_2$, allyl (Figure 4.3.1.7). A comparison of their relative abundances with those of the concomitantly formed LiMe$_6$Cu$_2$R$_2^-$ and LiMe$_6$Cu$_2$R'$_2^-$ species (i.e., **8**) shows an approximately statistical distribution and suggests that the reactions of **6** with the abovementioned substrates occur at similar rates.

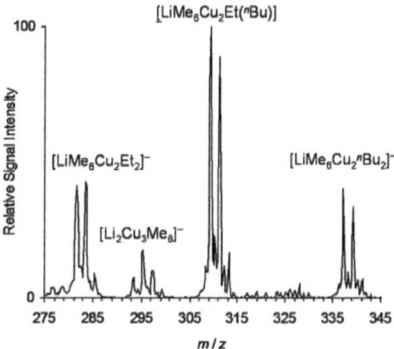

Figure 4.3.1.7. Section from the negative-ion mode ESI mass spectrum of a solution of the products formed in the reaction of $LiCuMe_2 \cdot LiCN$ with $EtI/^nBuI$ in THF.

4 Results and Discussion

4.3.2 Association Equilibria of Lithium Tetraalkylcuprates

Calculated Structures and Relative Energies in the Gas Phase. Cu(III) species adopt $3d^8$ valence electron configurations, for which square-planar coordination geometries are energetically most favorable. Such square-planar geometries have indeed been found for tetraalkylcuprate anions,[44b,45a,c,74] and theoretical calculations on **7a-f** fully confirm this result. The gas-phase calculations moreover suggest that the triple ions **8** contain two subunits of intact **7**, which each interact with Li^+ via two of their methyl groups to form a distorted tetrahedral coordination environment (4-Me coordination of Li^+, Figure 4.3.2.1; for the case of **8a**, test calculations with various theoretical methods consistently found similar coordination geometries).

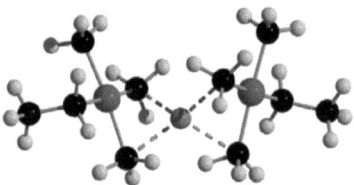

Figure 4.3.2.1. Calculated minimum energy structure of **8b** in the gas phase (grey: Cu, light grey: Li, black: C, white: H, B3LYP/6-31G*/SDD).

In contrast, involvement of the R groups in the Li^+ coordination is predicted to be energetically slightly less favorable (Table 4.3.2.1). This difference presumably results from the smaller size of the methyl substituents, which permits their closer approach to the Li^+ center (Table 4.3.2.1) and thereby enhances the electrostatic interaction. The preferential interaction of Li^+ with methyl groups has also been inferred above for cuprates(I) (Section 4.1.5).

Table 4.3.2.1. Relative energies (in kJ mol^{-1}) and Li-C bond distances (in pm) of the different isomers of **8a-f** according to B3LYP/6-31G*/SDD calculations

	4Me coordination of Li$^+$			3Me-R coordination of Li$^+$				2Me-2R coordination of Li$^+$			
	E_{rel}	r(Li-C$_{Me}$)^a	r(Li-Cu)^a	E_{rel}	r(Li-C$_{Me}$)^a	r(Li-C$_R$)	r(Li-Cu)^a	E_{rel}	r(Li-C$_{Me}$)^a	r(Li-C$_R$)^a	r(Li-Cu)^a
8a	0	225	250								
8b	0	224	251	6.4	224	231	249	12.1	224	234	247
8c	0	224	252	5.5	224	232	249	11.8	223	234	247
8d	0	224	251	5.7	224	233	248	11.6	223	234	248
8e	0	223	251	8.3	223	234	249	15.9	222	235	247
8f	0	223	251	7.0	223	233	249	15.8	221	238	247

^{a} Values given refer to the average of the different individual bond lengths.

Besides binding to the methyl groups, the Li$^+$ center may possibly also interact with the Cu atoms, given that the calculated Li-Cu distances are relatively short (Table 4.3.2.1) and that other d^8 systems, such as Pt(II),[97] have been shown to coordinate to Lewis acids via their d_{z^2} orbital. Presumably, the higher oxidation state of the Cu(III) atom substantially decreases the Lewis-basic character of its d_{z^2} orbital, however. In line with this argument, natural bond orbital analyses of the optimized structures of **8a** consistently find only rather weak Cu-Li interactions (Table 3.3.2).

For the case of **8a**, its dissociation energy according to Eq 4.3.2.1 with R = Me was also calculated. In the gas phase, this reaction is highly endothermic ($\Delta_{react}E$ = 165 kJ mol^{-1}). In solution, however, the situation most likely will be different because the release of the LiMe$_3$CuR moiety should be facilitated by solvation.

$$\text{LiMe}_6\text{Cu}_2\text{R}_2^- \rightarrow \text{Me}_3\text{CuR}^- + \text{LiMe}_3\text{CuR} \quad (4.3.2.1.)$$

Concentration- and Solvent-Dependent ESI Mass Spectrometric Measurements.
The formation of triple ions AB$_2^-$ from contact ion pairs A$^+$B$^-$ and free ions B$^-$ in solution is a well-known phenomenon.[98] Accordingly, concentration- and solvent-dependent measurements of mixtures of **6** and allyl choride RCl (CuCN/3 MeLi/RCl) were performed to gain further insight into the association equilibria leading to the formation of **8f**, Eq 4.3.2.2.

$$\text{Li}^+(\text{solv}) + 2\,\text{Me}_3\text{CuR}^- \rightleftharpoons \text{Li}^+\text{Me}_3\text{CuR}^-(\text{solv}) + \text{Me}_3\text{CuR}^- \rightleftharpoons \text{LiMe}_6\text{Cu}_2\text{R}_2^- \quad (4.3.2.2.)$$

With increasing concentration of CuCN/3 MeLi/RCl in THF, the relative ESI signal intensity of **7f** strongly decreases, whereas that of the triple ion **8f** rises correspondingly (Figure 4.3.2.2). This trend matches the behavior expected on the basis of the law of mass action, which predicts a shift toward higher aggregation states as a function of concentration (Eq 4.3.2.2).

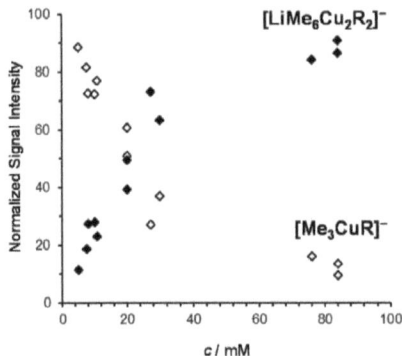

Figure 4.3.2.2. Normalized ESI signal intensities of Me_3CuR^- (open symbols) and of the corresponding triple ion $LiMe_6Cu_2R_2^-$ (closed symbols, R = allyl) as functions of the concentration c of CuCN/3 MeLi/RCl in THF.

For assessing the effect of the solvent, mixtures of CuCN/3 MeLi/RCl in CPME and MTBE were probed. Cu(III) species are not observed for reaction assays in the pure solvents, but only for solutions of Li^+**7f/8f** prepared by pre-formation in THF and further dilution. In the case of the CPME/THF mixtures, the observed fraction of monomeric **7f** is slightly decreased in comparison to the situation in pure THF (Table 4.3.2.2). This finding can be rationalized by the lower polarity and smaller Li^+ affinity of CPME, which make solvation less favorable and thus shift the equilibrium toward contact ion pairs and higher aggregation states (Eq 4.3.2.2). For mixtures of the even less polar MTBE with THF, one would expect a somewhat stronger effect, whereas just the opposite holds true (Table 4.3.2.2). Possibly, the interaction of MTBE molecules with the lithium cuprates(III) is so weak that they are displaced by THF molecules, thus giving rise to a local environment similar to that in pure THF solutions.

Table 4.3.2.2. Fraction of the monomeric cuprate(III) complex **2f** observed upon ESI of ethereal solutions (c = 10 mmol L^{-1}) of different polarity.

Solvent	Relative permittivity ε(298 K)	Fraction of monomeric **2f** in ESI mass spectrum[a]		
		Pure solvent	Solvent/THF 9:1	Solvent/THF 3:1
THF	7.42[b]	75 ± 5		
CPME	4.76[c]		50 ± 5	62 ± 5
MTBE	2.60[c]		75 ± 5	72 ± 5

[a] Defined as $I(\mathbf{2f})/[I(\mathbf{2f}) + I(\mathbf{3f})]$. [b] Reference 87. [c] Reference 88.

According to the above measurements, all probed lithium tetraalkylcuprates Li$^+$**7** have roughly similar tendencies to form the corresponding triple ions **8**. With ESI mass spectrometry, the absolute equilibrium concentrations of **8** in THF solutions cannot be determined due to the inherent limitations of the method (see Section 3.1.2). For NMR spectroscopy, in turn, the concentration of the triple ions **8** may be too low for their detection. Moreover, the interconversion between **7** and **8** could occur faster than the NMR time scale, which might explain why no triple ions **8** have been observed by this method.

Electrical Conductivity Studies. The molar electrical conductivity of **6** in THF is similar to those of the related lithium diorganylcuprates LiCuR'$_2$·LiCN, R' = ^{n}Bu, ^{t}Bu, and Ph. Based on a comparison of the measured conductivities with their estimated limiting conductivities (Section 4.1.7), it has been suggested that these reagents are not fully dissociated in THF, but partly form contact ion pairs. A similar situation can also be inferred for **6**. Upon the addition of 1 equiv of allyl chloride and the formation of the Li$^+$Me$_2$CuR(CN)$^-$ intermediate (R = allyl), the electrical conductivity significantly decreases (Figures 4.3.1.1 and 4.3.1.2). This decrease points to a lower dissociation tendency of Li$^+$Me$_2$CuR(CN)$^-$. Lithium cuprates(I) that contain cyanide ligands attached to the copper exhibit an analogous behavior, which is ascribed to the ambident nature of the cyanide ion and its ability to coordinate to copper and lithium centers simultaneously. The electrical conductivity of Li$^+$**7f** is higher again and roughly equals that of **6** (Figures 4.3.1.1 and 4.3.1.2), indicating a similar equilibrium between solvent-separated and contact ion pairs. The presence of the latter is a prerequisite for the formation of **8** according to Eq 4.3.2.2.

4.3.3. Unimolecular Reactivity of Tetraalkylcuprates

Fragmentation of Mononuclear Tetraalkylcuprate Anions. In the final step in the generally accepted mechanism of copper-mediated cross-coupling reactions, the Cu(III) intermediate releases the coupling product in a reductive elimination. Gas-phase experiments on tetraalkylcuprates as Cu(III) model systems offer the possibility to study this important elementary reaction in great detail. For mass-selected **7a**, the collision-induced dissociation (CID), as expected, leads to the formation of $CuMe_2^-$ and the concomitant elimination of ethane. Analogous experiments with labeled $Me_3CuCD_3^-$ determine the secondary kinetic isotope effect of this reaction as KIE = 1.0 ± 0.1 (determined for an excitation voltage of V_{exc} = 0.33 V). For the other, unsymmetrical tetraalkylcuprate anions **7b-g**, two different fragmentation channels are available: elimination of the cross-coupling product MeR or of the homo-coupling product ethane, Eq 4.3.3.1 and 4.3.3.2, respectively (Figure 4.3.3.1). If the homo-coupling MeCuR⁻ fragment ions contain β-hydrogen atoms, β-H eliminations can ensue and lead to MeCuH⁻ secondary fragment ions, as has already been shown above (Section 4.2.1 and Ref. 38d).

$Me_3CuR^- \rightarrow CuMe_2^- + MeR$ (4.3.3.1.)

$Me_3CuR^- \rightarrow MeCuR^- + Me_2$ (4.3.3.2.)

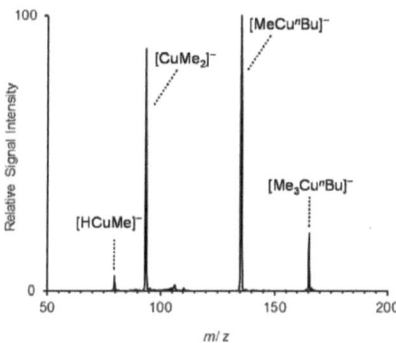

Figure 4.3.3.1. Mass spectrum of mass-selected **7d** (m/z = 165) and its fragment ions produced upon collision-induced dissociation (V_{exc} = 0.23 V).

For a comparison of the competition between cross- and homo-coupling reactions of the tetraalkylcuprates **7b-g**, relatively harsh CID conditions (V_{exc} = 0.25 – 0.30 V) are at first considered, resulting in the fragmentation of > 70% of the parent ion population. Whereas **7b** and **7c** preferentially afford the cross-coupling product, **7d** gives a 1:1 fragment ratio, which corresponds to a purely statistical branching; in contrast, **7e-g** mainly yield the homo-coupling product (Table 4.3.3.1). The fragmentation pattern of **7g** can also be compared with that of the related $Me_{4-n}CuR_n^-$ ions, R = $CF_3CH_2CH_2$ and n = 2 – 4. As expected, CuR_4^- only releases R_2, while $MeCuR_3^-$ exclusively eliminates MeR; $Me_2CuR_2^-$, in turn, loses MeR and Me_2 in approximately equal amounts (Table 4.3.3.2). The latter case is particularly interesting because $Me_2CuR_2^-$ can form two different isomers. For the *trans*-isomer, cross-coupling reactions should be strongly preferred, whereas the *cis*-isomer could yield both cross-coupling and homo-coupling products. The observed branching ratio suggests that the *cis*-isomer is at least partly present. Furthermore, the $CF_3CH_2CH_2$ group apparently has an intrinsically lower tendency to participate in the reductive elimination than methyl. For the related $Me_2Cu^nBu_2^-$ anion, the simultaneous occurrence of cross- and homo-coupling reactions (losses of $^nBu_2$ and Me_2) also points to the partial presence of the *cis*-isomer (Table 4.3.3.2).

Table 4.3.3.1. Branching fractions and appearance voltages V_{appear} (as approximate measures for relative threshold energies) of the fragmentation reactions of tetraalkylcuprate anions **7**.

Me_3CuR^- (R)	cross-coupling (Eq 4.3.3.1)		homo-coupling (Eq 4.3.3.2)	
	fraction[a]	V_{appear} [V]	fraction[a]	V_{appear} [V]
7a (Me)	0		1	
7b (Et)	0.93 ± 0.01	0.201 ± 0.001	0.07 ± 0.01	0.205 ± 0.005
7c (^nPr)	0.61 ± 0.04	0.214 ± 0.002	0.39 ± 0.04	0.215 ± 0.003
7d (^nBu)	0.50 ± 0.05	0.180 ± 0.002	0.50 ± 0.05	0.183 ± 0.002
7e (CH_2CH_2Ph)	0.06 ± 0.01	0.196 ± 0.005	0.94 ± 0.01	0.192 ± 0.001
7f ($CH_2CH=CH_2$)	0.21 ± 0.02	0.199 ± 0.003	0.79 ± 0.02	0.186 ± 0.002
7g ($CH_2CH_2CF_3$)	0.00 ± 0.00		1.00 ± 0.00	

[a] Determined for V_{exc} = 0.30 V.

4 Results and Discussion

Table 4.3.3.2. Branching ratios of the fragmentation reactions of further organocopper(III) anions.[a]

Parent ion	Fraction of Me$_2$ loss	Fraction of MeR loss	Fraction of R$_2$ loss
Me$_2$Cu(CH$_2$CH$_2$CF$_3$)$_2^-$	0.43 ± 0.03	0.57 ± 0.03	0.00 ± 0.00
MeCu(CH$_2$CH$_2$CF$_3$)$_3^-$	–	1.00	0.00
Me$_2$Cu^{n}Bu$_2^-$	0.03 ± 0.01	0.74 ± 0.04	0.24 ± 0.03

[a] Determined for V_{exc} = 0.30 V.

Upon variation of the excitation voltage V_{exc}, and, thus, the effective temperature of the parent ions **7a-f**, the branching ratios between cross- and homo-coupling reactions remain largely unchanged (Figure 4.3.3.2). Moreover, the determined appearance voltages V_{appear} for the fragmentations, corresponding to very approximate relative threshold energies (see Figure 3.1.4.1 for technical details), show a similar trend: the cross-coupling channel is energetically slightly favored for **7b-d**, whereas the homo-coupling channel is energetically more favorable for **7e** and **7f** (Table 4.3.3.1). This consistency indicates that the observed branching ratios reflect the true intrinsic behavior of the tetraalkylcuprate anions. Accordingly, a comparison of the present gas-phase data with solution-phase results from the literature appears meaningful.

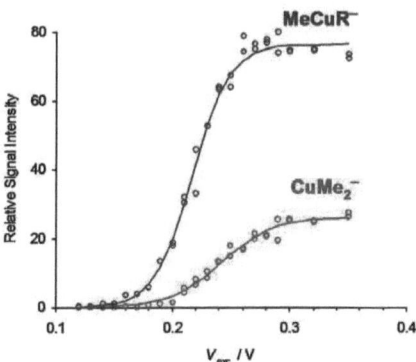

Figure 4.3.3.2. Fragment yields upon CID of mass-selected Me$_3$CuR$^-$ (R = allyl) as functions of V_{exc} together with sigmoid fits (see ref. 70b for further details).

In solution, reactions of **6** with simple aliphatic alkyl halides give the synthetically desired cross-coupling products in high yields,[41a] whereas increased amounts of homo-coupling were reported for a few reactions involving diallyl-[99] and dihexylcuprates.[100] In their theoretical analysis of the competition between cross-coupling and homo-coupling, Bäckvall, Nakamura, and coworkers focused on the coordination geometry of the neutral R'$_2$CuR intermediates for explaining the usually observed preference for cross-coupling reactions (Scheme 4.3.3.1).

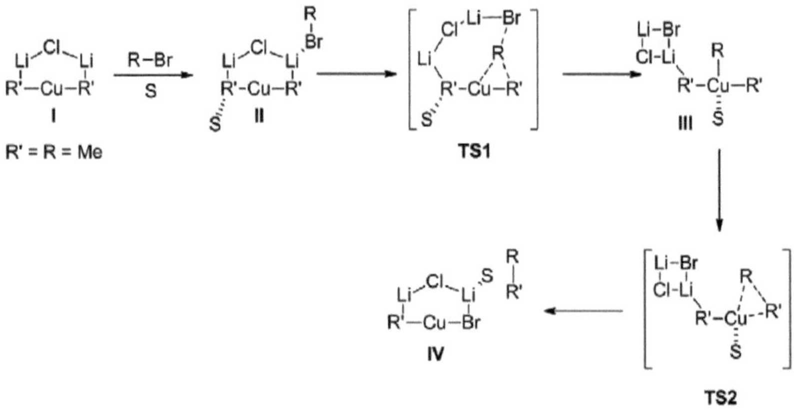

Scheme 4.3.3.1. Calculated reaction pathway for S$_N$2 alkylation reaction between LiCuMe$_2 \cdot$LiCl and MeBr in presence of Me$_2$O as solvent.

According to these calculations, the selective formation of the R–R' cross-coupling product is attributed to the trans relationship of the R' groups in the cuprate clusters **I** and **II**, which is preserved in **TS1** and in the Cu(III) intermediate **III**, leading to a T-shaped structure. Since reductive elimination is possible only for groups *cis* to one another, only cross-coupling takes place. The present results suggest that different organyl substituents may also have intrinsically different tendencies toward cross- or homo-coupling, respectively.

Calculated Fragmentation Pathways of Mononuclear Tetraalkylcuprate Anions.
Theory predicts high exothermicities for the fragmentation reactions of tetraalkylcuprate anions **7**, pointing to the low thermodynamic stability of the Cu(III) species (Table 4.3.3.3). Although DFT (B3LYP/6-31G*/SDD) and MP2 calculations with a larger basis set (MP2/6-311+G*/MDF) give considerably deviating *absolute* $\Delta_{react}E$ values, they find similar trends for the two competing fragmentation channels: While the cross-coupling reaction is significantly

more exothermic than the homo-coupling reaction for **7b** and moderately more exothermic for **7c** and **7d**, **7e** is a borderline case: the B3LYP calculations find a slightly larger exothermicity for the cross-coupling channel, whereas the MP2 calculations predict the homo-coupling reaction to be more exothermic. In contrast, both theoretical methods agree that the cross-coupling reaction is much less exothermic for **7f**. This reduced exothermicity can be largely ascribed to the weakness of the newly formed C–C bond in the cross-coupling product 1-butene (compared to the C-C bonds in saturated *n*-alkanes, such as the homo-coupling product Me_2 or the cross-coupling products formed from **7b-d**).

The theoretical activation energies $\Delta_{act}E$ display a parallel trend in that the relative preference for the homo-coupling reaction increases in the series **7b-f** (Table 4.3.3.3). Whereas the DFT calculations in all cases find a lower energy barrier for the homo-coupling reaction, the MP2 calculations with the larger basis set give more balanced barriers for both fragmentation channels. A comparison with the experimental V_{appear} values suggests that the calculations predict the correct trend for the activation barriers of the tetraalkylcuprates **7b-f**, but that they are biased in favor of the homo-coupling reaction. Free activation energies $\Delta_{act}G$ computed for a large temperature range of 298 ≤ T ≤ 1000 K do not vary significantly, which is in agreement with the small temperature sensitivity found experimentally.

Table 4.3.3.3. Calculated reaction and activation energies (in kJ mol^{-1}) of the fragmentation reactions of tetraalkylcuprate anions **7**.

Me_3CuR^- (R)	cross-coupling (Eq 4.3.3.1)				homo-coupling (Eq 4.3.3.2)			
	$\Delta_{react}E$		$\Delta_{act}E$		$\Delta_{react}E$		$\Delta_{act}E$	
	DFT^a	MP2^b	DFT^a	MP2^b	DFT^a	MP2^b	DFT^a	MP2^b
7a (Me)					−174.1	−147.4	139.6	120.9
7b (Et)	−181.2	−149.6	148.9	131.1	−167.9	−130.1	141.4	131.4
7c (Pr)	−177.4	−142.1	152.6	134.6	−170.2	−131.7	138.7	128.7
7d (^{n}Bu)	−177.7	−139.7	152.9	134.8	−170.9	−131.7	138.3	128.7
7e (CH_2CH_2Ph)	−173.2	−122.3	153.2	136.4	−172.5	−132.0	134.9	126.1
7f ($CH_2CH=CH_2$)	−144.0	−104.8	150.9	143.8^c	−177.4	−132.1	118.0	109.7

^{a} B3LYP/6-31G*/SDD. ^{b} MP2/6-311+G*/MDF. ^{c} Calculated for a geometry with the distance between the β-C atom of the allyl substituent and the Cu center held constant at 250 pm.

4 Results and Discussion

The calculated activation energies for the reductive elimination of the tetraalkylcuprates **7** are much larger than those for neutral trialkylcopper(III) species,[19a,43] for which values of $20 \leq E_{act} \leq 85$ kJ mol^{-1} have been predicted.[101,102] Of course, the higher kinetic stabilities of the former are a prerequisite for their successful detection in the present experiments. To understand the reason for the stabilization of the tetraalkylcuprate anions **7** at a qualitative level, the effect of attaching a methide ion to CuMe$_3$(III) on the one hand is compared to that of CuMe(I) on the other. In the case of the highly electron-deficient and Lewis-acidic copper(III) species CuMe$_3$, a large stabilization should result (Scheme 4.3.3.2). In contrast, the stabilization gained for CuMe is supposed to be much smaller. For the transition structure associated with the reductive elimination of Me$_2$, an intermediate behavior is expected, which corresponds to the inferred increased activation energy for the anionic copper(III) species.

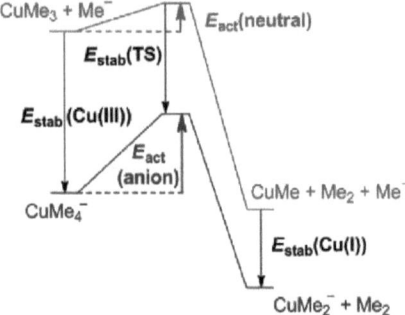

Scheme 4.3.3.2. Schematic potential energy surfaces for the reductive elimination of Me$_2$ from neutral CuMe$_3$ and anionic CuMe$_4^-$.

The reductive elimination of Me$_2$ from **7a** can also be compared with the analogous reaction of Me$_3$CuCl$^-$, which Pratt et al. have recently studied theoretically.[103] Both reactions exhibit similar transition structures of distorted tetrahedral geometries (Figure 4.3.3.3 for the reductive elimination of Me$_2$ from **7a**). However, the activation energy for the reductive elimination of Me$_2$ from **7a** is significantly higher. Presumably, this difference reflects the better stabilization of the Cu(III) center by a methide compared to that by a chloride ion (see above).

4   Results and Discussion

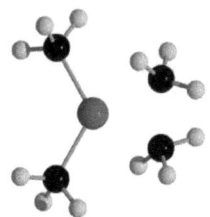

Figure 4.3.3.3. Calculated transition structure for the reductive elimination of Me$_2$ from **7a** (grey: Cu, black: C, white: H, B3LYP/6-31G*/SDD).

Fragmentation of LiMe$_6$Cu$_2$R$_2^-$ and LiMe$_6$Cu$_2$R(R')$^-$ Anions. The triple ions **8** form model systems that offer the possibility to assess the effect of a Li$^+$ counter-ion (paired with **7**) on the reactivity of the tetraalkylcuprates at a strictly molecular level. Like their monomeric counterparts **7**, the dimeric complexes **8** afford cross- and homo-coupling reactions upon CID, Eq 4.3.3.3 and 4.3.3.4, respectively (Figure 4.3.3.4). The resulting mixed cuprate(I/III) fragment ions easily undergo consecutive reductive eliminations to form LiCu$_2$Me$_4^-$. The latter is partly hydrolyzed by a reaction with background water present in the ion trap (Section 4.2.2).

$$\text{LiMe}_6\text{Cu}_2\text{R}_2^- \rightarrow \text{LiMe}_5\text{Cu}_2\text{R}^- + \text{MeR} \quad (4.3.3.3.)$$

$$\text{LiMe}_6\text{Cu}_2\text{R}_2^- \rightarrow \text{LiMe}_4\text{Cu}_2\text{R}_2^- + \text{Me}_2 \quad (4.3.3.4.)$$

No elimination of R$_2$ homo-coupling products was observed. Although their formation is clearly disfavored on simple statistical grounds, this argument appears insufficient to explain the complete absence of these reactions. Instead, this absence is interpreted as another indication of **8** being composed of two separate subunits **7**, in line with theoretical calculations (see Section 4.3.2).

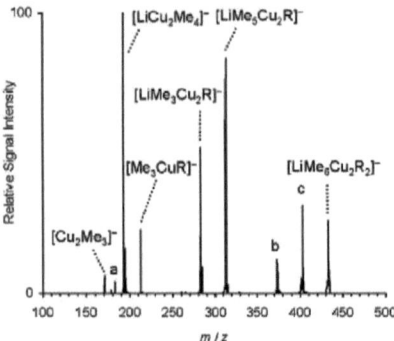

Figure 4.3.3.4. Mass spectrum of mass-selected LiMe$_6$Cu$_2$R$_2^-$ (m/z = 433, R = PhCH$_2$CH$_2$) and its fragment ions produced upon collision-induced dissociation (V_{exc} = 0.15 V), a = MeCuR$^-$, b = LiMe$_2$Cu$_2$R$_2^-$, c = LiMe$_4$Cu$_2$R$_2^-$. The ion at m/z = 195 corresponds to LiMe$_3$Cu$_2$(OH)$^-$, which results from an ion-molecule reaction of LiCu$_2$Me$_4^-$ (m/z = 193) with background water present in the ion trap.

Compared to the monomeric tetraalkylcuprates **7**, the presence of the additional Li$^+$**7** subunit in **8b-e** substantially enhances the fraction of the cross-coupling (Table 4.3.3.4). This difference is rationalized by the preferential interaction of the Li$^+$ ion with the methyl groups in **8** (see Section 4.3.2). With the four central methyl groups thus tied up, only the terminal Me and R substituents of **8** are supposedly prone to reductive elimination, thereby yielding the cross-coupling products. A deviating behavior is observed for **8f**·THF, which forms a rare example of an anionic THF complex sufficiently stable to survive the ESI process (detected for CPME/THF mixtures). Possibly, the presence of the solvent molecule in **8f**·THF changes the coordination geometry of its LiMe$_6$Cu$_2$R$_2^-$ core in such a way that it no longer favors the cross-coupling channel (Table 4.3.3.4).

Table 4.3.3.4. Branching fractions of the fragmentation reactions of the triple ions **8**[a, b].

LiMe$_6$Cu$_2$R$_2^-$ (R)	fraction of cross-coupling (Eq 4.3.3.3)	fraction of homo-coupling (Eq 4.3.3.4)
8b (Et)	0.93 ± 0.03	0.00 ± 0.00
8c (^{n}Pr)	0.91 ± 0.02	0.00 + 0.01
8d (^{n}Bu)	0.86 ± 0.05	0.00 ± 0.00
8e (CH$_2$CH$_2$Ph)	0.54 ± 0.04	0.10 ± 0.04
8f·THF (CH$_2$CH=CH$_2$)	0.07 ± 0.03^c	0.56 ± 0.03^c

[a] The given fractions do not add up to 1 because of the presence of fragment ions, such as Cu$_2$Me$_3^-$ and LiMe$_3$Cu$_2$R$^-$, which cannot be unambiguously assigned to cross- or homo-coupling as primary fragmentation reaction, respectively. [b] Determined for V_{exc} = 0.30 V. [c] Determined for V_{exc} = 0.22 V.

Fragmentation experiments on *mixed* triple ions LiMe$_6$Cu$_2$R(R')$^-$ moreover make possible a direct comparison of different substituents R/R' in their tendency to undergo reductive elimination. Such a comparison is particularly straightforward because the presence of the two R/R' groups in the same parent ion ensures the availability of equal amounts of energy for both fragmentation pathways. The measured branching ratios consistently point to a clear order in the intrinsic reactivity of the different organyl substituents (Figure 4.3.3.5), which also largely agrees with the trends inferred from the fragmentation experiments on the mononuclear species **8b-f** (the fact that for the latter, the homo-coupling fraction observed for **8e** exceeds that for **8f** seems to be an anomaly; note that the appearance voltages V_{appear} derived for the homo-coupling reactions of both species show the reversed order and are thus consistent with the behavior of the mixed triple ions).

4   Results and Discussion

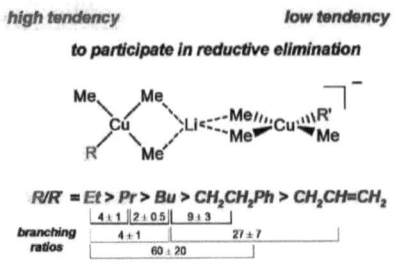

Figure 4.3.3.5. Tendencies of different R/R' groups toward reductive elimination, as determined from the fragmentation of mixed triple ions LiMe$_6$Cu$_2$R(R')$^-$. The given branching ratios (listed at the bottom) are based not only on the observed signal intensities of the primary fragment ions (corresponding to losses of MeR and MeR', respectively), but also take into account secondary fragmentation channels (losses of MeR/Me$_2$ and MeR'/Me$_2$, respectively).

Calculated Fragmentation Pathways of LiCu$_2$Me$_8^-$ and LiCu$_2$Me$_6^-$ Anions.
DFT calculations (B3LYP/6-31G*/SDD) show that the reductive elimination of Me$_2$ from **8a** ($\Delta_{react}E = -186.9$ kJ mol^{-1}) is more exothermic and has a smaller energy barrier ($\Delta_{act}E = 97.2$ kJ mol^{-1}) than the analogous reaction of **7a** (Table 4.3.3.3). Presumably, the central Li$^+$ ion weakens the Lewis basicity of the attached methyl groups and, thus, reduces their stabilizing effect on the Cu(III) centers (see above); analogous behavior may also be expected for the other **8** anions. Note that the calculated fragmentation pathway involves two *terminal* methyl groups (Figure 4.3.3.6), in line with the qualitative arguments raised above.

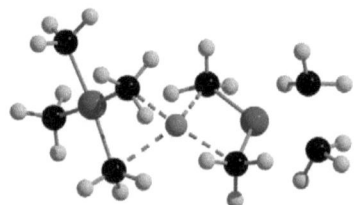

Figure 4.3.3.6. Calculated transition structure for the reductive elimination of Me$_2$ from **8a** (grey: Cu, light grey: Li, black: C, white: H, B3LYP/6-31G*/SDD).

4 Results and Discussion

The resulting primary fragment ion $LiCu_2Me_6^-$ consists of a square-planar $CuMe_4^-$ and a linear $CuMe_2^-$ subunit (Figure 4.3.3.7), which coordinates to the central Li^+ ion via a single methyl group. The consecutive reductive elimination of Me_2 from $LiCu_2Me_6^-$ ($\Delta_{react}E =$ −181.2, $\Delta_{act}E$ = 96.6 kJ mol^{-1}) yields the complex $LiCu_2Me_4^-$, which contains two linear $CuMe_2^-$ subunits (Figure 4.3.3.7).

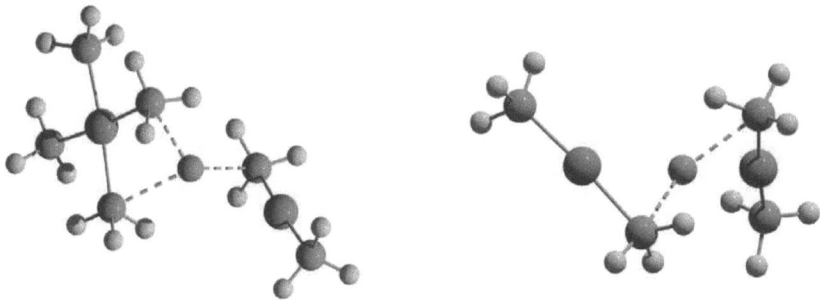

Figure 4.3.3.7. Predicted structures of the primary and secondary fragment ions $LiCu_2Me_6^-$ and $LiCu_2Me_4^-$ (B3LYP/6-31G*/SDD).

4 Results and Discussion

4.4. Conjugate Addition Reactions

4.4.1 Reactions of Diorganylcuprates with Acrylonitrile.

Upon addition of one equivalent of acrylonitrile to THF solutions of LiCuR$_2$·LiCN (R = Me, ^{n}Bu, Ph) a yellow color is observed, along with precipitate formation. No ionic products were observed on analysis of R = ^{n}Bu, Ph systems by negative-ion mode ESI mass spectrometry. However, for the more reactive Me system, a certain degree of organocuprate decomposition is observed, presumably due to a significant extent of addition (both 1,4 and 1,2) along with polymerization.

4.4.2 Reactions of Diorganylcuprates with Fumaronitrile.

Reactions of fumaronitrile (FN) with cyanocuprates in THF afford dark brown solutions, in which adducts [CuR$_2$·FN]$^-$ are detected for all R. These species show very limited macroscopic stability, which decreases in the row ^{n}Bu > Me > Ph. Related complexes of higher nuclearity are observed only for the case of R = ^{n}Bu (Figure 4.4.2.1). When the reaction with fumaronitrile is conducted in Et$_2$O, a significant shift towards higher aggregation states is observed (Figure 4.4.2.2).

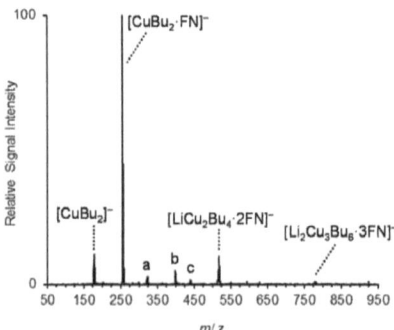

Figure 4.4.2.1. Negative-ion mode ESI-MS of a solution of the products formed in the reaction of LiCu^{n}Bu$_2$·LiCN with fumaronitrile (FN) in THF, a = LiCu$_2$^{n}Bu$_3$(OH)$^-$, b = [LiCu$_2$^{n}Bu$_3$(OH)·FN]$^-$, c = [LiCu$_2$^{n}Bu$_4$·FN]$^-$.

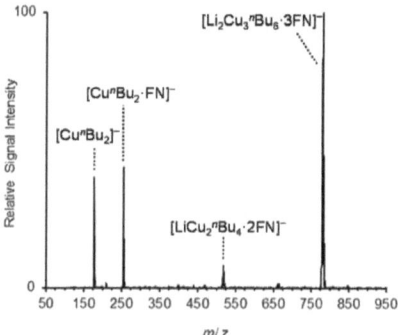

Figure 4.4.2.2. Negative-ion mode ESI-MS of a solution of the products formed in the reaction of LiCu^{n}Bu$_2$·LiCN with fumaronitrile (FN) in Et$_2$O.

Gas-phase CID experiments on the detected complexes help further clarify their structure. So, all of the complexes observed decompose by liberating intact fumaronitrile (Figures 4.4.2.3 – 4.4.2.5).

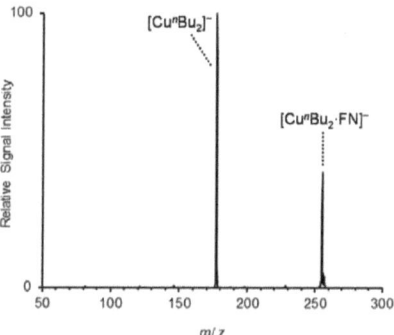

Figure 4.4.2.3. Mass spectrum of mass-selected [Cu^{n}Bu$_2$·FN]$^-$ (m/z = 255) and its fragment ion produced upon collision-induced dissociation (V_{exc} = 0.21 V).

The absence of other decomposition pathways supports the suggestion that these species are genuine π-complexes, and not isobaric insertion products (Scheme 2.4.1). In the latter case one might expect to see, among others, the formation of species with the organyl group attached to the fumaronitrile moiety. For polynuclear π-complexes, the occurrence of severe in-trap hydrolysis reactions (Figures 4.4.2.4 and 4.4.2.5) implies a rather open structure. These aggregates, of the formula $[Li_{x-1}(Cu^nBu_2 \cdot FN)_x]^-$, can be thought to be composed of monomeric $[Cu^nBu_2 \cdot FN]^-$ subunits and lithium cations. Alternatively, these species can be viewed as derivatives of the parent organocuprate anions $Li_{x-1}Cu_x^nBu_{2x}^-$, with the FN ligands coordinating to both copper (via C=C) and lithium (via C≡N).

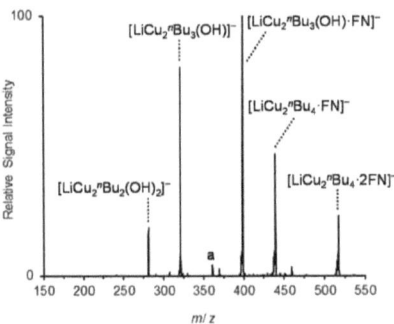

Figure 4.4.2.4. Mass spectrum of mass-selected $[LiCu_2^nBu_4 \cdot 2FN]^-$ (m/z = 255) and its fragment ions produced upon collision-induced dissociation (V_{exc} = 0.16 V), a = $LiCu_2^nBu_4^-$.

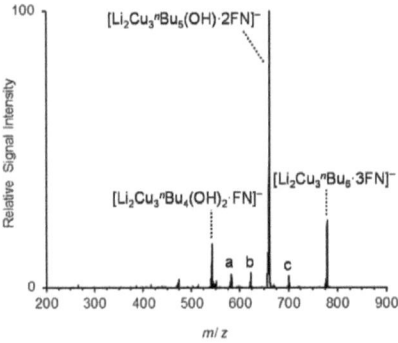

Figure 4.4.2.5. Mass spectrum of mass-selected $[Li_2Cu_3^nBu_6 \cdot 3FN]^-$ (m/z = 255) and its fragment ions produced upon collision-induced dissociation (V_{exc} = 0.23 V), a = $[Li_2Cu_3^nBu_5(OH) \cdot FN]^-$, b = $[Li_2Cu_3^nBu_6 \cdot FN]^-$, c = $[Li_2Cu_3^nBu_6 \cdot 2FN]^-$.

4 Results and Discussion

The formation of these larger aggregates in Et$_2$O as compared to THF is in line with the above results on homoleptic cyanocuprates (Section 4.1.6). It is interesting to note, however, that the corresponding butyl dimer LiCu$_2$^{n}Bu$_4^-$ was not detected in solutions of parent LiCu^{n}Bu$_2$·LiCN, in either THF or Et$_2$O. Possibly, the presence of extra fumaronitrile ligands in the dimer helps better hold it together.

4.4.3 Reactions of Diorganylcuprates with 1,1-dicyanoethylene.

When neat 1,1-dicyanoethylene is added to organocuprate solutions in THF, a bead of polymer is instantly formed. To overcome this difficulty, the substrate was added as a solution in THF, resulting in formation of green solutions. In this case no π-complexes are seen, but products of Michael addition, both in the cation (Figure 4.4.3.1) and anion modes.

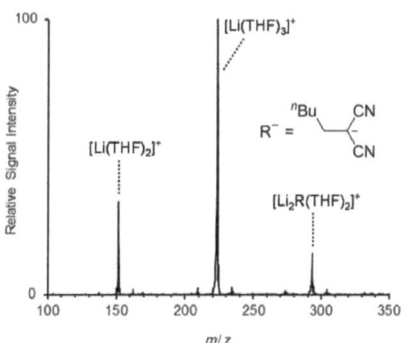

Figure 4.4.3.1. Positive-ion mode ESI-MS of a solution of the products formed in the reaction of LiCu^{n}Bu$_2$·LiCN with 1,1-dicyanoethylene in THF.

4.4.4 Reactions of Diorganylcuprates with Tricyanoethylene.

Reactions of tricyanoethylene with diorganylcuprates afford orange-brown solutions. Anions of the empiric formula HC$_4$(CN)$_5$R$^-$ are detected (R = Me, ^{n}Bu), both free and incorporated into cuprate structures (Figure 4.4.4.1).

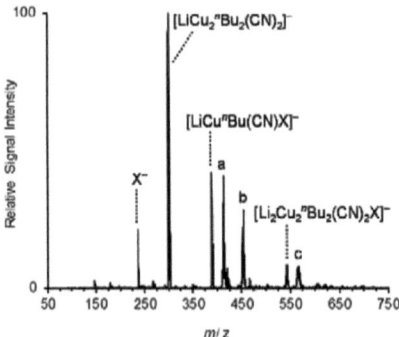

Figure 4.4.4.1. Negative-ion mode ESI-MS of a solution of the products formed in the reaction of LiCu^{n}Bu$_2$·LiCN with tricyanoethylene in THF, a = Li$_2$Cu$_3$^{n}Bu$_2$(OH)(CN)$_3^-$, b = Li$_2$Cu$_3$^{n}Bu$_3$(CN)$_3^-$, c = Li$_3$Cu$_4$^{n}Bu$_3$(OH)(CN)$_4^-$; X = HC$_4$(CN)$_5$^{n}Bu.

From the composition of these anions, a plausible explanation of their formation is the combination of two tricyanoethylene units with one R$^-$ anion, followed by elimination of HCN. One of the many mechanistic explanations possible is given in Scheme 4.4.4.1.

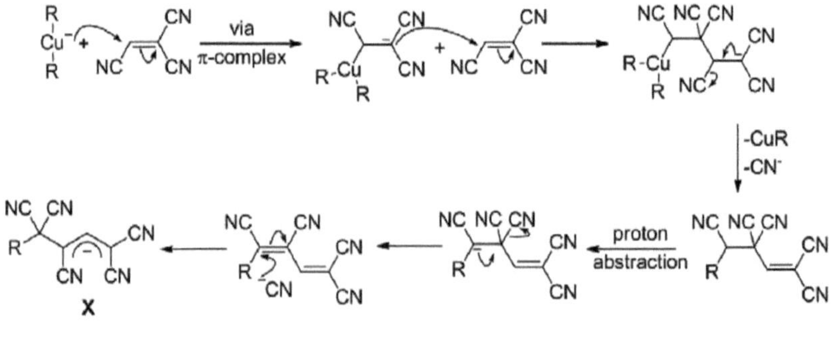

Scheme 4.4.4.1. Proposed mechanism of reaction between LiCuR$_2$·LiCN and tricyanoethylene, R = ^{n}Bu, Me.

When free, the abovementioned anions X$^-$ fragment by HCN loss in the gas-phase. However, upon incorporation into the cuprate structure, radical loss is observed for the case of R = ^{n}Bu

(Eq 4.4.4.1 and Figure 4.4.4.2; CID of analogous Me species not possible due to low signal intensity and short lifetime).

$$LiCu^nBu(CN)X^- \rightarrow LiCu(CN)X^- + {^nBu^\bullet} \qquad (4.4.4.1.)$$

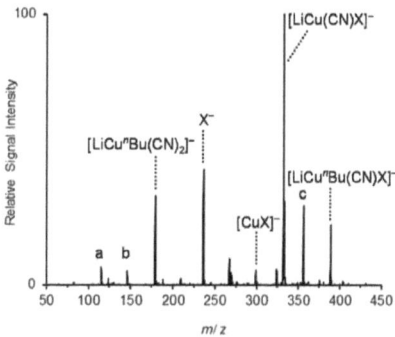

Figure 4.4.4.2. Mass spectrum of mass-selected [LiCu^{n}Bu(CN)X]$^-$ (m/z = 389) and its fragment ions produced upon collision-induced dissociation (V_{exc} = 0.24 V), a = Cu^n(CN)$_2^-$, b = Cu^{n}Bu(CN)$^-$, c = Cu^{n}Bu(X)$^-$; X = HC$_4$(CN)$_5$^{n}Bu. Note the loss of a butyl radical.

To provide additional evidence, an alternative experiment with dihexylcuprate (LiCuHex$_2 \cdot$LiCN) was conducted. Species analogous to the butyl case were detected, the CID experiments of which also showed radical loss. Possibly, a redox reaction takes place between the anions and Cu(I) centers present, resulting in formation of open-shell intermediates, which then fragment to lose butyl/hexyl radicals.

Diphenylcuprate behaves differently from its alkyl analogues, in that a different anion (of the composition C$_3$(CN)$_5^-$) is detected in its reactions with tricyanoethylene. The suggested mechanism of this transformation (Scheme 4.4.4.2) is thought to be closely related to that for R = ^{n}Bu and Me.

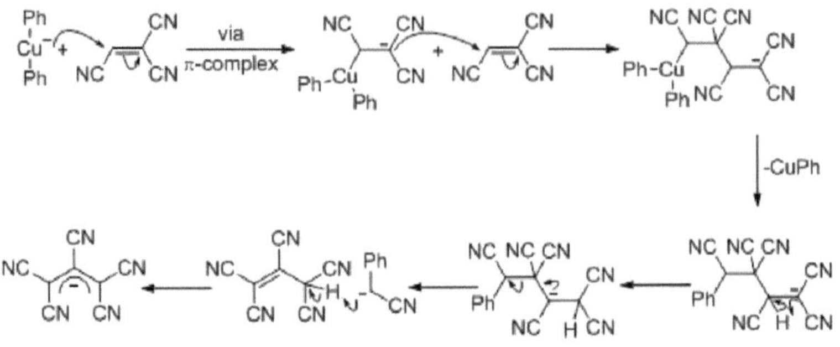

4 Results and Discussion

Scheme 4.4.4.2. Proposed mechanism of reaction between diphenylcuprate and tricyanoethylene.

In this case, however, isomerization of one of the anionic intermediates leads to an anion that can eliminate PhCH(CN)¯ (pK_a of conjugate acid ca. 22 in DMSO),[104] which is better stabilized than the corresponding Me and ^{n}Bu analogues (pK_a of conjugate acid ca. 32 in DMSO).[105] Subsequent proton transfer results in a very stable allylic anion, stabilized by five cyano groups.

4.4.5 Reactions of Diorganylcuprates with Tetracyanoethylene.

Tetracyanoethylene, the most potent π-acceptor of the series, reacts with organocuprates to generate light-yellow colored solutions with precipitate. Cu(III) intermediates can be observed in all cases, with the rationale of their formation given in Scheme 4.4.5.1.

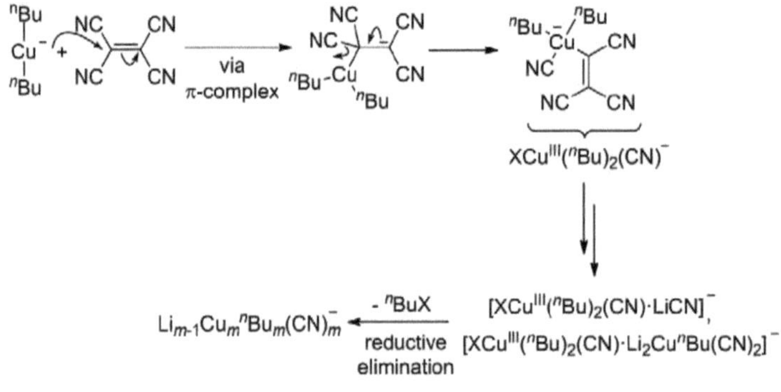

Scheme 4.4.5.1. Proposed mechanism of reaction between LiCuR$_2$·LiCN and tetracyanoethylene, exemplified by R = ^{n}Bu.

In the first step, the cuprate attacks the electron-poor double bond, possibly via an intermediate π-complex. The Cu(III) intermediate formed can rearrange to give an isobaric tetracoordinate anionic Cu(III) species, reminiscent of the tetraalkylcuprates previously detected and described above. This species can then form further aggregates in solution. A typical spectrum of the ^{n}Bu system is shown below (Figure 4.4.5.1).

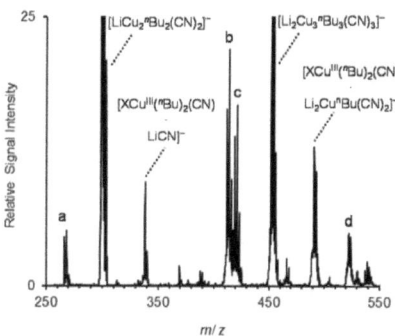

Figure 4.4.5.1. Negative-ion mode ESI-MS of a solution of the products formed in the reaction of LiCu^{n}Bu$_2$·LiCN with tetracyanoethylene in THF, zoomed in to the area of interest; a = Cu$_2$^{n}Bu$_2$CN$^-$, b = Li$_2$Cu$_3$^{n}Bu$_2$(OH)(CN)$_3$$^-$, c = LiCu$_3^{n}Bu_3(CN)_2$$^-$, d = XCu^{III}(^{n}Bu)$_2$(CN)·Li$_2$Cu^{n}Bu$_2CN^-$, X = C$_2(CN)_3$.

Proof of identity is given by the CIDs of the [XCu^{III}R$_2$(CN)·LiCN]$^-$ anions (R = Me, ^{n}Bu, Ph). They all lose RX upon fragmentation, whereas an isobaric tetracyanoethylene π-complex is expected to lose tetracyanoethylene (XCN), just as the corresponding fumaronitrile complexes lose fumaronitrile. A typical example for R = ^{n}Bu is given below (Figure 4.4.5.2)

4   Results and Discussion

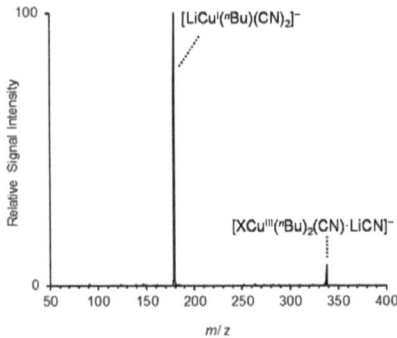

Figure 4.4.5.2. Mass spectrum of mass-selected $[\text{XCu}^{III}(^n\text{Bu})_2(\text{CN})\cdot\text{LiCN}]^-$ (m/z = 338) and its fragment ion produced upon collision-induced dissociation (V_{exc} = 0.25 V).

The structure of other Cu(III) species $[\text{XCu}^{III}\text{R}_2(\text{CN})\cdot\text{Li}_2\text{CuR}(\text{CN})_2]^-$ is not clear. Whereas species with R = Me, Ph decompose similarly to their mononuclear counterparts (by loss of RX, see Figure 4.4.5.3 and Eq 4.4.5.1), and hence can be assumed to have the same Cu(III) core, the corresponding butyl anion undergoes radical loss (Figure 3.4.5.4 and Eq 4.4.5.2).

$\text{XCu}^{III}\text{R}_2(\text{CN})\cdot\text{Li}_2\text{CuR}(\text{CN})_2^- \rightarrow \text{Li}_2\text{Cu}_2\text{R}_2(\text{CN})_3^- + \text{RX}$ (4.4.5.1.)

$\text{XCu}^{IIIn}\text{Bu}_2(\text{CN})\cdot\text{Li}_2\text{Cu}^n\text{Bu}(\text{CN})_2^- \rightarrow \text{Li}_2\text{Cu}_2{}^n\text{Bu}_2(\text{X})(\text{CN})_3^- + {}^n\text{Bu}^\bullet$ (4.4.5.2.)

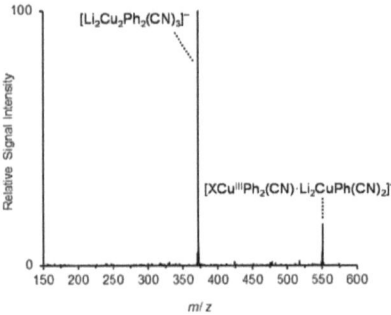

Figure 4.4.5.3. Mass spectrum of mass-selected $[\text{XCu}^{III}\text{Ph}_2(\text{CN})\cdot\text{Li}_2\text{CuPh}(\text{CN})_2]^-$ (m/z = 551) and its fragment ion produced upon collision-induced dissociation (V_{exc} = 0.20 V).

4 Results and Discussion

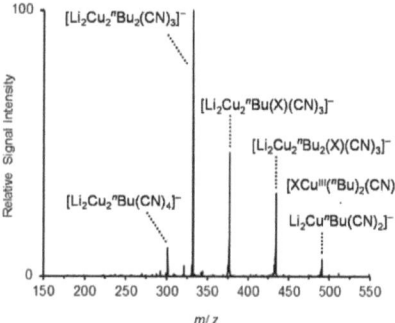

Figure 4.4.5.4. Mass spectrum of mass-selected $[XCu^{III}(^nBu)_2(CN)\cdot Li_2Cu^nBu(CN)_2]^-$ (m/z = 491) and its fragment ions produced upon collision-induced dissociation (V_{exc} = 0.20 V). Note the loss of a butyl radical.

To confirm that the observed pathway corresponds to radical loss, a hexyl system was studied, just as for the case of tricyanoethylene. The observed fragmentation indeed points to loss of free hexyl radicals, with the origins of this phenomenon probably lying in the formation of Cu(II) species or tetracyanoethylene-based radical anions.[106,107]

4.4.6 Substrate Structure-Reactivity Relationships.

The simplest Michael acceptor of the series, acrylonitrile, does not form detectable ionic products with $LiCuR_2 \cdot LiCN$ reagents. This is rationalized by the known ability of organocuprates to induce polymerization of acrylonitrile.[108] In this case, the cuprate is consumed in only catalytic amounts, hence no evidence for a reaction is observed by ESI mass spectrometry. Besides, the polarizing effect of the single cyano group on the C=C bond is probably too weak to allow the formation of stable π-complexes.

Fumaronitrile, with two cyano groups attached to the double bond, forms a range of π-complexes in different aggregation states. Their relative stability can be explained by the electronic structure of the substrate. Compared to acrylonitrile, the presence of a second electron-withdrawing CN group increases the acceptor potency of the double bond, which is expected to favor back-donation from the cuprate. The symmetry of the molecule means that the electron distribution is not skewed towards either end, which inhibits further reaction of the π-complex formed (cf. 1,1-dicyanoethylene). Moreover, the electron-deficient C=C bond reduces the importance of a stabilizing lithium-heteroatom interaction, as demonstrated by the detection of Li-free [CuR$_2$·FN]$^-$. In contrast, this interaction between the lithium of the cuprate aggregate and the heteroatom of the coordinated Michael acceptor is crucial for α,β-unsaturated carbonyls. Indeed, for their case, no Li-free π-complexes have ever been observed or suggested.

As opposed to fumaronitrile, in the case of 1,1-dicyanoethylene the two cyano groups polarize the double bond in the *same* direction, making conversion of the π-complex into the Cu(III) intermediate more favorable. Once formed, this highly reactive species then undergoes rapid reductive elimination that produces the corresponding product anion.

For tri- and tetrasubstituted cyanoethylenes, the presence of a leaving group at the site of cuprate attack, together with the combined electron-withdrawing effect of three or four cyano groups opens the possibility for new reaction pathways, resulting in rich chemistry observed.

5 Conclusions and Outlook

The present work explores the aggregation state, structure and reactivities of organocuprate intermediates, which can be broadly divided into two classes. The first is represented by homo- and heteroleptic cyanocuprates(I), LiCuR$_2$·LiCN and LiCuR(CN), respectively. Copper(III) intermediates of cross-coupling and conjugate addition reactions, together with related π-complexes, represent the second class. In both cases, ESI mass spectrometry in ethereal solvents permits the detection and characterization of a wide range of non-stabilized anionic intermediates, as well as providing insight into association equilibria they undergo. Although the ionization process most likely shifts these equilibria relative to the situation in solution, the qualitative trends derived seem to be remarkably robust.

So, for LiCuR$_2$·LiCN solutions, the cyanide-free anions Li$_{n-1}$Cu$_n$R$_{2n}^-$, $n = 1$ and 3, along with Li$_2$(CN)(solv)$_n^+$ cations are the predominant species observed. The abovementioned anionic aggregates are in equilibrium with each other, the position of this equilibrium being strongly solvent dependent. Thus, non-polar solvents (e.g. Et$_2$O) favor higher aggregation states, whereas monomeric CuR$_2^-$ dominate in the more polar THF. Based on fragmentation experiments, the trimeric Li$_2$Cu$_3$R$_6^-$ complexes were shown to correspond to adducts of anionic CuR$_2^-$ and the neutral dimeric contact ion pair Li$_2$Cu$_2$R$_4$, i.e. **1**. Previous studies have indeed identified exactly these species as the major constituents of LiCuR$_2$·LiCN reagents in ethereal solvents. However, whereas presence of simple Li(solv)$_n^+$ cations was inferred in those investigations, dinuclear Li$_2$(CN)(solv)$_n^+$ cations are observed by ESI mass spectrometry. In this respect, the present results more closely agree with previous IR and X-ray absorption spectroscopic measurements, which also suggested the prominence of the Li$_2$(CN)$^+$ motif, although tied up in the contact ion pair **2**. While not fully reconciling the two conflicting views put forward in the literature, the present findings thus indicate that both of them capture important aspects of LiCuR$_2$·LiCN reagents in THF and other ethereal solvents.

In addition to the influence of solvent on aggregation, ESI-MS makes possible a direct comparison of the substituent effects and shows that their increased steric hindrance results in higher degrees of dissociation, whose absolute values can be estimated on the basis of the measured molar conductivities. On the contrary, dissociation is hindered by the introduction of the small polar CN groups, which can bridge between copper centers and strongly coordinate to Li cations. So, polynuclear Li$_{n-1}$Cu$_n$R$_n$(CN)$_n^-$ anions were detected in solutions

5 Conclusions and Outlook

of LiCuR(CN), with nuclearities much higher than those of the cyanide-free $\text{Li}_{n-1}\text{Cu}_n\text{R}_{2n}^-$ anions.

Following the determination of factors influencing the organocuprate aggregation state, its importance for reactivity is demonstrated. So, gas-phase hydrolysis of cuprate ions shows, *inter alia*, that the presence of Li^+ can significantly enhance the reaction rate, e.g., CuR_2^- monomers do not react with water under the experimental conditions, whereas the corresponding $\text{LiCu}_2\text{R}_4^-$ dimers and $\text{Li}_2\text{Cu}_3\text{R}_6^-$ trimers do.

The aggregation state is also of key importance for tetraalkylcuprates(III), detected in cross-coupling reactions of organocuprates with alkyl iodides. Monomeric tetraalkylcuprate anions **7** (Me_3CuR^-) show variable amounts of cross- and homo-coupling in their fragmentation reactions, depending on the nature of the R group. The triple ions **8** ($\text{LiMe}_6\text{Cu}_2\text{R}_2^-$), which consist of two subunits of **7** held together by a Li cation, however, very strongly favor cross-coupling. The theoretical calculations reveal that the origin of this selectivity is the preferential interaction of the central lithium cation with two methyl groups of each subunit, which thereby blocks the homo-coupling pathway. Besides highlighting the importance of aggregation state in organocuprate chemistry, the observation of tetraalkylcuprates is valuable for other reasons as well. While so far almost exclusively neutral organocopper(III) species have been considered as intermediates in copper-mediated cross-coupling reactions, the present experiments suggest that the participation of tetraalkylcuprate anions should also be taken into account if the overall reagent stoichiometry allows their formation (3 equiv of Me/R anions per Cu atom). This findings opens exciting prospects for further harnessing the unique reactivity of Cu(III) in organic synthesis.

The mild nature of the ESI process makes possible the detection of species with weak, essentially non-covalent interactions holding them together, such as organocuprate π-complexes with Michael acceptors. Here, it has been demonstrated that these highly fluxional species show complex aggregation equilibria, the position of which is influenced by solvent polarity in a way similar to that of the parent organocuprates. The stability of the abovementioned π-complexes depends on the electronic structure of the substrate. Interactions that are too weak do not result in detectable complexes, too strong interaction make the reaction proceed further and yield either Cu(III) intermediates or final conjugate addition products.

5 Conclusions and Outlook

To sum up, the successful detection and characterization of the abovementioned organocopper species shows the tremendous potential of ESI mass spectrometry for the analysis of charged organometallics. Put into perspective, the present work adds to a growing number of studies that demonstrate the suitability of ESI mass spectrometry for monitoring reactions and probing ion speciation in solution. Until this finding has been further validated, however, the most reliable approach, adopted in this work, remains the combination of ESI mass spectrometry with other, well-established analytical methods, such as electrical conductivity measurements or NMR spectroscopy.

Outlook. With regard to the type of organocuprates(I), a possible extension of the present work might probe CuX salts other than CuCN (X = halogen, SCN, OTf, RC≡C, PhS, etc.) and establish the influence of the counter-anion on solution-phase composition. Probing systems of intermediate stoichiometry, e.g. the well known Ashby's cuprates $LiCu_2Me_3$ and $Li_2Cu_3Me_5$,[109] represents another possible direction of future work. The range of solvents used in all of the experiments might also be extended to those of non-ethereal nature, such as CH_2Cl_2 or CH_3CN.

The bimolecular gas-phase reactivity of organocuprate anions, so far probed only in reactions with water, should further be extended to reactions with synthetically more useful electrophiles, such as alkyl halides and Michael acceptors. In this way, the currently held opinion that only homodimers undergo conjugate addition can be directly put to test.

Another direction of further efforts may deal with extending the range of Me_3CuR^- species detected to their aryl analogues, Me_3CuAr^-, and establishing a synthetic procedure for the general preparation of $[R^1{}_3CuR^2]^-$, $R^1 \neq Me$ in solution. The gas-phase reductive eliminations of those ions will increase the size of the present database and allow further insights into the cross-coupling process. Furthermore, the Cu intermediates are not limited to cross-couplings and conjugate additions, but have been proposed for many other reactions, like the copper-catalyzed Ullmann reaction[110], the recently reported site-selective arylation of arenes[111], and the Suzuki coupling of 2-pyridyl substrates[112]. Detecting the organocopper intermediates involved and shedding light on the catalytic cycle of these reactions would allow catalyst optimization and development of further synthetically useful Cu-catalyzed processes.

Finally, π-complexes of standard α,β-unsaturated carbonyl electrophiles and the related Cu(III) intermediates have so far remained elusive to ESI-MS and await detection and

5 Conclusions and Outlook

characterization. The ESI-MS approach could also be extended to elucidate the mechanism of alkyne and diene carbocuprations, which is believed to involve π-complexes as well, but remains unclear. Studies of these systems might clarify some aspects of this synthetically useful transformation.

6 Appendix

6.1. Analytical Data

Nuclear Magnetic Resonance Spectroscopy. NMR spectra were recorded on Varian Mercury 200 and Bruker AC 300 instruments. Chemical shifts are reported as δ-values in ppm relative to the solvent signal. For the characterization of the observed signal multiplicities the following abbreviations are used: s (singlet), d (doublet), t (triplet), q (quartet), and br (broad).

UV/VIS Spectroscopy. UV/VIS spectra were recorded from 200 to 800 nm on a Perkin Elmer Lambda 16 spectrometer. Samples were measured as solutions in Et_2O, and the absorption bands reported in nm.

Infrared Spectroscopy. Infrared spectra were recorded from 400 to 4000 cm^{-1} on a Perkin 281 IR spectrometer. Samples were measured neat (ATR, Smiths Detection DuraSample IR II Diamond ATR). The absorption bands were reported in wave numbers (cm^{-1}).

6.2. Synthesis

6.2.1 General Considerations

Standard Schlenk techniques were employed for handling air- and moisture-sensitive substances throughout. THF and Et$_2$O were distilled from sodium/benzophenone; MeTHF, CPME, and MTBE were dried over molecular sieve (4Å). CuCN was dried by repeated heating under vacuum at 350 °C. Solutions of organolithium compounds RLi were used as purchased: MeLi (1.49 M) in Et$_2$O, EtLi (0.42 M) in benzene/cyclohexane (90/10), ^{n}BuLi (2.37 M) in hexane, ^{s}BuLi (1.58 M) in cyclohexane, ^{t}BuLi (1.88 M) in pentane, and PhLi (1.74 M) in ^{n}Bu$_2$O. The exact concentrations were determined by titration of 1,3-diphenyl-2-propanone tosylhydrazone.[113]

For the labeling experiments, CD$_3$I (Sigma Aldrich, 99.5% D content), EtI-D$_5$ (Sigma Aldrich, 99.5% D content), BuI-D$_9$ (Ehrenstorfer, 99.3% D content) and K^{13}CN (Sigma Aldrich, 99.0 % ^{13}C content) were employed.

Other solvents and chemicals were used as purchased.

6.2.2 Synthesis of Organocuprate Reagents

Conductivity Measurements. To prepare ethereal solutions of LiCuR$_2$·LiCN and LiCuR(CN) (R = Me, ^{n}Bu, ^{t}Bu, and Ph), CuCN was heated in an evacuated flask at 350 °C for five minutes. The flask was then filled with Ar and cooled to RT, and the procedure was repeated two more times. The screw cap of the flask was removed, and the electrode (pre-dried by heating at 120 °C for 3 minutes) was put in. A suspension of CuCN in the solvent of choice (THF, Et$_2$O) was then treated with one or two equivalents of RLi under argon at −15 °C. After stirring at this temperature for 15 minutes, the CuCN completely dissolved, forming LiCuR(CN) or LiCuR$_2$·LiCN, respectively. Solutions of LiCuMe$_2$·LiCN in THF (of nominal concentration c = 100 mM) with and without added allyl chloride were analyzed at −71 °C in order to slow down interfering hydrolysis reactions. Nonetheless, the latter were found to reduce the concentration of the active dimethylcuprate reagent by 20 ± 5% (as determined by iodometric titration). The amount of hydrolysis thus exceeds that determined above for LiCuR$_2$·LiCN solutions (R = ^{n}Bu, ^{t}Bu, and Ph), which points to a particularly high sensitivity of the dimethylcuprate reagent.

ESI-MS Probes. CuCN was heated in an evacuated Schlenk flask at 350 °C for three minutes. The flask was then filled with Ar and cooled to RT, and the procedure was repeated two more times. Solutions of CuCN/(RLi)$_m$ (R = Me, Et, ^{n}Bu, ^{s}Bu, ^{t}Bu, and Ph) were prepared by treating suspensions of CuCN in the solvent of choice (THF, Et$_2$O, MeTHF, CPME, or MTBE) with RLi under argon at −78 °C. After stirring at this temperature for 1 h, the CuCN completely dissolved for CuCN/(RLi)$_m$, m = 2 or 1, forming LiCuR$_2$·LiCN and LiCuR(CN), respectively. In the case of the sample solutions prepared for ESI mass-spectrometric analysis, the smaller volumes necessarily increase the likelihood of small errors in the measured reagent quantities. Such errors can be particularly detrimental to the analysis of LiCuR(CN). So, for m = 1, signals typical of LiCuR$_2$·LiCN systems were dominant in some cases, which was ascribed to addition of a slight excess of RLi. To avoid this, test experiments with m = 0.8 and 0.5 equivalents of RLi were conducted for THF systems. It was found that the intensity of the abovementioned homoleptic peaks was significantly lower in these cases. The resulting solutions of nominal Li$_m$CuR$_m$(CN) composition are supposed to contain LiCuR(CN) because the excess CuCN does not dissolve, as could also be directly seen from the presence of a solid residue. On this basis, m = 0.8 systems were chosen to study LiCuR(CN) speciation in Et$_2$O.

Aliquots of the resulting solutions (of typical concentrations c = 25 mM) were then transferred into a gastight syringe and introduced into the ESI source of a mass spectrometer. Sample solutions of LiCuR$_2$·LiCN showed relatively high macroscopic stabilities in the syringe held at room temperature. In contrast, solutions of Li$_m$CuR$_m$(CN), $m \leq 1$, decomposed in ≤ 10 min and produced black or greenish precipitates, which then caused clogging of the inlet line connecting the syringe with the ESI source. To avoid this problem, sample solutions of Li$_m$CuR$_m$(CN), $m \leq 1$, had to be analyzed as quickly as possible.

Sample solutions of LiCuMe$_2$·LiCN/RX stoichiometry (R = Me, Et, ^{n}Pr, ^{n}Bu, PhCH$_2$CH$_2$, CH$_2$=CHCH$_2$, and CF$_3$CH$_2$CH$_2$) were prepared by adding MeLi (2 equiv) to suspensions of CuCN in dry ethereal solvents at −78 °C and stirring at this temperature for 15 min to approx. 1 h, before the organyl halide RX was added (1 equiv). Addition of further MeLi (1 equiv) afforded sample solutions of CuCN/3 MeLi/RX stoichiometry, which alternatively could be prepared by treatment of CuCN suspensions with 3 equiv of MeLi (−78 °C, 1 h), followed by the addition of RX (1 equiv). Analogous procedures provided solutions of LiCuR'$_2$·LiCN/RX. Sample solutions of LiCuR$_2$·LiCN/C$_2$H$_{4-n}$(CN)$_n$ stoichiometry (R = Me, ^{n}Bu and Ph), were prepared by adding RLi (2 equiv) to suspensions of CuCN in dry ethereal solvents at −78 °C and stirring at this temperature for 15 min to approx. 1 h, before the corresponding cyanoethylene Michael acceptor C$_2$H$_{4-n}$(CN)$_n$ was added (1 equiv).

6.2.3 Synthesis of Cu^{13}CN[114]

K^{13}CN (250 mg, 3.85 mmol) was dissolved in water (8 mL) at room temperature. A solution of Na$_2$SO$_3$ (256 mg 2.03 mmol) in water (8 mL) was added, followed by a solution of CuSO$_4$·5H$_2$O (1.0 g, 4 mmol) in H$_2$O (8 mL). The resulting colorless precipitate was stirred for 10 minutes, and a solution of NaOH (70 mg, 1.75 mmol) in H$_2$O (2 mL) was added. After 20 minutes the solid was allowed to settle and the liquor was decanted. The resulting product was washed with water (2×10 mL), acetone (3×10 mL) and dried *in vacuo* to give Cu^{13}CN as colorless solid (305 mg, 3.40 mmol, 88%).

6.2.4 Synthesis of Cyanoethylene Substrates

1,1-Dicyanoethylene[115]

A dry 50-mL round-bottom flask was charged with 1,1,3,3-tetracyanopropane (10.0 g, 70 mmol) and fitted with a U-tube cooled in a water bath ($\theta \approx 15$ °C). A small amount of P_2O_5 was placed over the crystals and inside the U-tube to avoid polymerization of product. After evacuation to 5 mbar, the system was heated (temperature gradient 180 to 250 °C) for 30 minutes to give 6 mL of clear liquid. The collected product was then purified by fractional distillation at 5 mbar. Fraction 1 (4 mL) was collected at 55 °C, while a second fraction (2 mL) was collected at 75 °C. Spectral data indicated that fraction 1 was pure 1,1-dicyanoethylene (3.8 g, 49 mmol, 70%). ^{1}H NMR (200 MHz, CDCl$_3$): δ 6.8 (s, 2H).

Ethyl 2,3-Dicyanopropionate[116]

A mixture of sodium cyanide (9.8 g, 0.20 mol), ethylcyanoacetate (22.6 g, 0.20 mol) and paraformaldehyde (6.0 g, 0.2 mol) was stirred in absolute ethanol (200 mL) for 10 minutes and then refluxed for 30 minutes. The solution was allowed to cool, then poured into a mixture of hydrochloric acid (0.5 M, 400 mL) and CH$_2$Cl$_2$ (300 mL). The layers were separated, and the aqueous phase extracted with CH$_2$Cl$_2$ (2×300 mL). The organic layers were combined, dried and concentrated *in vacuo* to afford the crude product as reddish oil (26 g). Vacuum distillation (5·10^{-3} mbar) afforded pure ethyl 2,3-dicyanopropionate as a colorless oil (19.2 g, 0.13 mol, 63%), bp 120 °C. ^{1}H NMR (200 MHz, CDCl$_3$): δ 4.36 (q, $J = 7$ Hz, 2H), 3.86 (t, $J = 7$ Hz, 1H), 3.01 (d, $J = 7$ Hz, 2H), 1.37 (t, $J = 7$ Hz, 3H).

2,3-Dicyanopropionamide[116]

NC–CH(CN)–CONH$_2$

A mixture of ethyl 2,3-dicyanopropionate (19.2 g, 0.13 mol) and concentrated aqueous ammonia (36 mL) was stirred for 4 hours at 0 °C. The colorless precipitate was filtered, washed with water and dried *in vacuo* to give the product as a colorless solid (13.2 g, 0.11 mol, 85%). ^{1}H NMR (200 MHz, DMSO-d_6): δ 7.9 (br s, 1H), 7.7 (br s, 1H), 4.15 (t, J = 7 Hz, 1H), 3.10 (d, J = 7 Hz, 2H).

1,1,2-Tricyanoethane[116]

NC–CH(CN)–CH$_2$–CN

2,3-dicyanopropionamide (13.2 g, 0.11 mol) was mixed with sodium chloride (18.0 g), acetonitrile (55 mL) and POCl$_3$ (9.8 mL) and the mixture stirred at room temperature for 5 minutes, followed by a 5-hour reflux. The resulting suspension was filtered and the solid residue washed with acetonitrile. The filtrate was concentrated *in vacuo* to ca. 20 mL and water was added. The precipitate was filtered at 0 °C, washed with water and dried *in vacuo* to afford the product as light-purple crystals (8.82 g, 84 mmol, 78%). ^{1}H NMR (200 MHz, DMSO-d_6): δ 5.29 (t, J = 6 Hz, 1H), 3.59 (d, J = 6 Hz, 2H).

1-Bromo-1,1,2-tricyanoethane[116]

(NC)$_2$C(Br)–CH$_2$–CN

A suspension of 1,1,2-tricyanoethane (4.0 g, 38 mmol) in water (40 mL) was cooled in an ice bath and bromine (2.0 mL, 39 mmol) was dropwise added, the temperature was kept below 6 °C. Ten minutes after the addition was complete, the mixture was filtered, and the moist solid dissolved in CH$_2$Cl$_2$. The solution was dried and concentrated *in vacuo* to afford the product as a colorless solid (6.0 g, 33 mmol, 86%). ^{1}H NMR (200 MHz, CDCl$_3$): δ 3.56 (s, 1H).

6 Appendix

Tricyanoethylene[116]

$$\text{NC}-\text{CH}=\text{C}(\text{CN})_2$$

1-Bromo-1,1,2-tricyanoethane (6.0 g, 33 mmol) was dissolved in Et$_2$O (30 mL) and a solution of triethylamine (3.1 g, 31 mmol) in Et$_2$O (15 mL) was added dropwise at 0 °C. The mixture was filtered and the solid washed with Et$_2$O. The filtrate was concentrated *in vacuo* to give crude product as dark oil (2.0 g, 19.4 mmol), attempted purification of which on silica gel failed. The crude tricyanoethylene was recovered (1.6 g, 8.2 mmol) and suspended in isohexane (10 mL). Diethyl ether was dropwise added to achieve full dissolution. The solution was cooled in a dry ice-acetone bath and the liquid phase decanted. The orange crystalline solid was dried *in vacuo* to afford pure tricyanoethylene (200 mg, 1.9 mmol, 6%). ^{1}H NMR (300 MHz, CDCl$_3$): δ 6.80 (s, 1H); ^{13}C NMR (300 MHz, CDCl$_3$): δ 128.0, 111.4, 109.7, 108.9, 105.2; UV: λ_{max} = 237 nm, ε_{max} = 12700; IR(neat): ν_{max} 3052, 2246, 2198, 1587, 1502, 1323, 1166, 1024, 998, 778 cm^{-1}.

The analytical data obtained are in agreement with literature values:[116]
UV: λ_{max} = 237 nm, ε_{max} = 13100;
IR(neat): ν_{max} 3030, 2222, 1502 cm^{-1}.

6.3. Determination of Background Water Concentration

To estimate the concentration of background water, calibration reactions with known values of the true second-order rate constant k_2 were run. Hydrolysis of magnesium acetylides, $RC{\equiv}CMgCl_2^-$, R = H and Ph, described in detail by O'Hair et al,[117] was chosen as reference. Under the experimental conditions, these reactions also showed the expected pseudo first order kinetics (Table 6.3.1).

Table 6.3.1. Gas-phase hydrolysis rate constants of magnesium acetylides $RCCMgCl_2^-$.

Parent ion	k_1/s^{-1}	k_2^a / molecule$^{-1}\cdot$cm$^3\cdot$s^{-1}	[H$_2$O] / molecule$\cdot$cm^{-3}
$HCCMgCl_2^-$	20 ± 0.4	$0.22 \cdot 10^{-9}$	$9.3 \cdot 10^{10}$
$PhCCMgCl_2^-$	17 ± 2	$0.27 \cdot 10^{-9}$	$6.3 \cdot 10^{10}$

^{a} Errors were conservatively estimated as ± 25% by the authors.

7 References and Notes

[1] *The Chemistry of Organocopper Compounds*; Rappoport, Z.; Marek, I., Eds.; Wiley: Hoboken, 2009.

[2] Boettger, R. *Liebigs Ann. Chem.* **1859**, *109*, 351-352.

[3] Buckton, G. *Liebigs Ann. Chem.* **1859**, *109*, 218-227.

[4] Reich, R. *Seances Acad. Sci.* **1923**, *177*, 322-324

[5] Gilman, H.; Straley, J. M. *Recl. Trav. Chim. Pays-Bas* **1936**, *55*, 821-834

[6] Gilman, H.; Jones, R. G.; Woods, L. A. *J. Org. Chem.* **1952**, *17*, 1630-1634.

[7] Corey, E. J.; Posner, G. H. *J. Am. Chem. Soc.* **1967**, *89*, 3911-3912.

[8] House, H. O.; Respess, W. L.; Whitesides, G. M. *J. Org. Chem.* **1966**, *31*, 3128-3141.

[9] Whitesides, G. M.; Fisher, W. F.; San Filippo, J.; Bashe, R. W.; House, H. O. *J. Am. Chem. Soc.* **1969**, *91*, 4871-4882.

[10] G. H. Posner, C. E. Whitten, J. J. Sterling, *J. Am. Chem. Soc.* **1973**, *95*, 7788-7800.

[11] Normant, J. F. *Pure & Appl. Chem.*, **1978**, *50*, 709—715.

[12] (a) Gorlier, J.-P.; Hamon, L.; Levisalles, J.; Wagnon, J. *J. Chem. Soc., Chem. Commun.* **1973**, 88. (b) Mandeville, W. H.; Whitesides, G. M. *J. Org. Chem.* **1974**, *39*, 400-405. (c) Koosha, K.; Berlan, J.; Capmau, M.-L.; Chodkiewicz, W. *Bull. Soc. Chim. France*, **1975**, 1284-1290. (d) Acker, R.-D. *Tetrahedron Lett.* **1977**, *18*, 3407-3410. (e) Four, P.; Riviere, H.; Tang, P. W. *Tetrahedron Lett.* **1977**, *18*, 3879-3882.

[13] Knochel, P.; Yeh, M. C. P.; Berk, S. C.; Talbert, J. *J. Org. Chem.* **1988**, *53*, 2392-2394.

[14] (a) Lipshutz, B. H.; Wilhelm, R. S.; Kozlowski, J. A. *Tetrahedron* **1984**, *40*, 5005-5038. (b) Lipshutz, B. H. *Synthesis* **1987**, 325-341.

[15] Krause, N.; Gerold, A. *Angew. Chem.* **1997**, *109*, 194-213; *Angew. Chem. Int. Ed. Engl.* **1997**, *36*, 186-204.

[16] (a) Polet, D.; Alexakis, A. In *The Chemistry of Organocopper Compounds*; Rappoport, Z.; Marek, I., Eds.; Wiley: Hoboken, 2009, pp 693-730. (b) Krause, N.; Aksin-Artok, Ö. In *The Chemistry of Organocopper Compounds*; Rappoport, Z.; Marek, I., Eds.; Wiley: Hoboken, 2009, pp 857-879.

[17] Chemla, F.; Ferreira, F. In *The Chemistry of Organocopper Compounds*; Rappoport, Z.; Marek, I., Eds.; Wiley: Hoboken, 2009, pp 527-584.

[18] Spino, C. In *The Chemistry of Organocopper Compounds*; Rappoport, Z.; Marek, I., Eds.; Wiley: Hoboken, 2009, pp 603-691.

[19] (a) Nakamura, E.; Mori, S. *Angew. Chem.* **2000**, *112*, 3902-3924; *Angew. Chem. Int. Ed.* **2000**, *39*, 3750-3771. (b) Nakamura, E.; Yoshikai, N. In *The Chemistry of Organocopper Compounds*; Rappoport, Z.; Marek, I., Eds.; Wiley: Hoboken, 2009, pp 1-21.

[20] (a) Gschwind, R. M. *Chem. Rev.* **2008**, *108*, 3029-3053. (b) Gärtner, T.; Gschwind, R. M. In *The Chemistry of Organocopper Compounds*; Rappoport, Z.; Marek, I., Eds.; Wiley: Hoboken, 2009, pp 163-215.

[21] Van Koten, G.; Jastrzebski, J. T. B. H. In *The Chemistry of Organocopper Compounds*; Rappoport, Z.; Marek, I., Eds.; Wiley: Hoboken, 2009, pp 23-143.

[22] Woodward, S. *Chem. Soc. Rev.* **2000**, *29*, 393-401

7 References and Notes

[23] (a) Lipshutz, B. H.; Wilhelm, R. S.; Floyd, D. M. *J. Am. Chem. Soc.* **1981**, *103*, 7672-7674. (b) Lipshutz, B. *Synlett* **1990**, 119-128. (c) Lipshutz, B. H.; Sharma, S.; Ellsworth, E. L. *J. Am. Chem. Soc.* **1990**, *112*, 4032-4034. (d) Lipshutz, B. H.; James, B. *J. Org. Chem.* **1994**, *59*, 7585-7587.

[24] (a) Bertz, S. H. *J. Am. Chem. Soc.* **1990**, *112*, 4031-4032. (b) Snyder, J. P.; Bertz, S. H. *J. Org. Chem.* **1995**, *60*, 4312-4313. (c) Bertz, S. H.; Miao, G.; Eriksson, M. *Chem. Commun.* **1996**, 815-816.

[25] (a) Stemmler, T.; Penner-Hahn, J. E.; Knochel, P. *J. Am. Chem. Soc.* **1993**, *115*, 348-350. (b) Snyder, J. P.; Spangler, D. P.; Behling, J. R. *J. Org. Chem.* **1994**, *59*, 2665-2667. (c) Barnhart, T. M.; Huang, H.; Penner-Hahn, J. E. *J. Org. Chem.* **1995**, *60*, 4310. (d) Stemmler, T. L.; Barnhart, T. M.; Penner-Hahn, J. E.; Tucker, C. E.; Knochel, P.; Böhme, M.; Frenking, G. *J. Am. Chem. Soc.* **1995**, *117*, 12489-12497. (e) Mobley, T. A.; Müller, F.; Berger, S. *J. Am. Chem. Soc.* **1998**, *120*, 1333-1334.

[26] Boche, G.; Bosold, F.; Marsch, M.; Harms, K. *Angew. Chem.* **1998**, *110*, 1779-1781; *Angew. Chem. Int. Ed.* **1998**, *37*, 1684-1686.

[27] Kronenburg, C. M. P.; Jastrzebski, J. T. B. H.; Spek, A. L.; van Koten, G. *J. Am. Chem. Soc.* **1998**, *120*, 9688-9689.

[28] Krause, N. *Angew. Chem.* **1999**, *111*, 83-85; *Angew. Chem. Int. Ed.* **1999**, *38*, 79-81.

[29] (a) Gschwind, R. M.; Rajamohanan, P. R.; John, M.; Boche, G. *Organometallics* **2000**, *19*, 2868-2873. (b) John, M.; Auel, C.; Behrens, C.; Marsch, M.; Harms, K.; Bosold, F.; Gschwind, R. M.; Rajamohanan, P. R.; Boche, G. *Chem. Eur. J.* **2000**, *6*, 3060-3068. (c) Gschwind, R. M.; Xie, X.; Rajamohanan, P. R.; Auel, C.; Boche, G. *J. Am. Chem. Soc.* **2001**, *123*, 7299-7304.

[30] (a) Xie, X.; Auel, C.; Henze, W.; Gschwind, R. M. *J. Am. Chem. Soc.* **2003**, *125*, 1595-1601. (b) Henze, W.; Vyater, A.; Krause, N.; Gschwind, R. M. *J. Am. Chem. Soc.* **2005**, *127*, 17335-17342.

[31] Huang, H.; Alvarez, K.; Cui, Q.; Barnhart, T. M.; Snyder, J. P.; Penner-Hahn, J. E. *J. Am. Chem. Soc.* **1996**, *118*, 8808-8816; correction: **1996**, *118*, 12252.

[32] Huang, H.; Liang, C. H.; Penner-Hahn, J. E. *Angew. Chem.* **1998**, *110*, 1628-1630; *Angew. Chem. Int. Ed.* **1998**, *37*, 1564-1566.

[33] Gerold, A.; Jastrzebski, J. T. B. H.; Kronenburg, C. M. P.; Krause, N.; van Koten, G. *Angew. Chem.* **1997**, *109*, 778-780; *Angew. Chem. Int. Ed. Engl.* **1997**, *36*, 755-757.

[34] Bertz, S. H.; Nilsson, K.; Davidsson, Ö.; Snyder, J. P. *Angew. Chem.* **1998**, *110*, 327-331; *Angew. Chem. Int. Ed.* **1998**, *37*, 314-317.

[35] (a) Bertz, S. H.; Eriksson, M.; Miao, G.; Snyder, J. P. *J. Am. Chem. Soc.* **1996**, *118*, 10906-10907. (b) Bertz, S. H.; Chopra, A.; Ogle, C. A.; Seagle, P. *Chem. Eur. J.* **1999**, *5*, 2680-2691.

[36] (a) Krauss, S. R.; Smith, S. G. *J. Am. Chem. Soc.* **1981**, *103*, 141-148. (b) Nakamura, E.; Mori, S.; Morokuma, K. *J. Am. Chem. Soc.* **1997**, *119*, 4900-4910. (c) Mori, S.; Nakamura, E. *Chem. Eur. J.* **1999**, *5*, 1534-1543.

[37] (a) Lipshutz, B. H.; Stevens, K. L.; James, B.; Pavlovich, J. G.; Snyder, J. P. *J. Am. Chem. Soc.* **1996**, *118*, 6796-6797. (b) Lipshutz, B. H.; Keith, J.; Buzard, D. J. *Organometallics* **1999**, *18*, 1571-1574.

[38] a) James, P. F.; O'Hair, R. A. *J. Org. Lett.* **2004**, *6*, 2761-2764. (b) Rijs, N.; Khairallah, G. N.; Waters, T.; O'Hair, R. A. *J. Am. Chem. Soc.* **2008**, *130*, 1069-1079. (c) Rijs, N. J.; Yates, B. F.; O'Hair, R. A. *J. Chem. Eur. J.* **2010**, *16*, 2674-2678. (d) Rijs, N. J.; Yoshikai, N.; Nakamura, E.; O'Hair, R. A. *J. Am. Chem. Soc.* **2012**, *134*, 2569-2580.

[39] For the few cases of known stable organocopper(II) species, see: Van Koten, G.; Jastrzebski, J. T. B. H. In *The Chemistry of Organocopper Compounds*; Rappoport, Z.; Marek, I., Eds.; Wiley: Hoboken, 2009, pp 23-143.

7 References and Notes

[40] For the very few cases of reported stable organocopper(III) species, see: (a) Willert-Porada, M. A.; Burton, D. J.; Baenziger, N. C. *J. Chem. Soc., Chem. Commun.* **1989**, 1633-1634. (b) Naumann, D.; Roy, T.; Tebbe, K.-F.; Crump, W. *Angew. Chem.* **1993**, *105*, 1555-1556; *Angew. Chem. Int. Ed. Engl.* **1993**, *32*, 1482-1483.

[41] (a) Whitesides, G. M.; Fischer, W. F., Jr.; San Filippo, J., Jr.; Bashe, R. W.; House, H. O. *J. Am. Chem. Soc.* **1969**, *97*, 4871-4882. (b) Vermeer, P.; Meijer, J.; Brandsma, L. *Recl. Trav. Chim.* **1975**, *94*, 112-114. (c) Luche, J. L.; Barreiro, J. M. Dollat, Crabbé, P. *Tetrahedron Lett.* **1975**, *16*, 4615-4618. (d) Goering, H. L.; Kantner, S. S. *J. Org. Chem.* **1983**, *48*, 721-724.

[42] Nakamura, E.; Yamanaka, M. *J. Am. Chem. Soc.* **1999**, *121*, 8941-8942.

[43] Snyder, J. P. *Angew. Chem.* **1995**, *107*, 112-113; *Angew. Chem. Int. Ed. Engl.* **1995**, *34*, 80-81.

[44] (a) Bertz, S. H.; Cope, S.; Murphy, M.; Ogle, C. A.; Taylor, B. J. *J. Am. Chem. Soc.* **2007**, *129*, 7208-7209. (b) Bertz, S. H.; Cope, S.; Dorton, D.; Murphy, M.; Ogle, C. A. *Angew. Chem.* **2007**, *119*, 7212-7215; *Angew. Chem. Int. Ed.* **2007**, *46*, 7082-7085. (c) Bartholomew, E. R.; Bertz, S. H.; Cope, S.; Dorton, D. C.; Murphy, M.; Ogle, C. A. *Chem. Commun.* **2008**, 1176-1177. (d) Bartholomew, E. R.; Bertz, S. H.; Cope, S.; Murphy, M.; Ogle, C. A. *J. Am. Chem. Soc.* **2008**, *130*, 11244-11245. (e) Bartholomew, E. R.; Bertz, S. H.; Cope, S. K.; Murphy, M. D.; Ogle, C. A.; Thomas, A. A. *Chem. Commun.* **2010**, *46*, 1253-1254. (f) Bertz, S. H.; Murphy, M. D.; Ogle, C. A.; Thomas, A. A. *Chem. Commun.* **2010**, *46*, 1255-1256. (g) Bertz, S. H.; Moazami, Y.; Murphy, M. D.; Ogle, C. A.; Richter, J. D.; Thomas, A. A. *J. Am. Chem. Soc.* **2010**, *132*, 9549-9551.

[45] (a) Gärtner, T.; Henze, W.; Gschwind, R. M. *J. Am. Chem. Soc.* **2007**, *129*, 11362-11363. (b) Henze, W.; Gärtner, T.; Gschwind, R. M. *J. Am. Chem. Soc.* **2008**, *130*, 13718-13726. (c) Gärtner, T.; Yoshikai, N.; Neumeier, M.; Nakamura, E.; Gschwind, R. M. *Chem. Commun.* **2010**, *46*, 4625-4626.

[46] Perlmutter, P. Conjugate Addition Reactions in Organic Synthesis; Pergamon: New York, 1992; Chapter 1.

[47] (a) Bertz, S.H.; Carlin, C. M.; Deadwyler, D. A.; Murphy, M. D.; Ogle, C. A.; Seagle, P. *J. Am. Chem. Soc.* **2002**, *124*, 13650-13651. (b) Murphy, M. D.; Ogle, C. A.; Bertz, S. H. *Chem. Commun.* **2005**, 854-856. (c) Bertz, S. H.; Hardin, R. A.; Murphy, M. D.; Ogle, C. A.; Richter, J. D.; Thomas, A. A. *Angew. Chem., Int. Ed.* **2012**, *51*, 2681-2685.

[48] (a) Hallnemo, G.; Olsson, T.; Ullenius, C. *J. Organomet. Chem.* **1985**, *282*, 133-144. (b) Lindstedt, E. L.; Nilsson, M.; Olsson, T. *J. Organomet. Chem.* **1987**, *334*, 255-261. (c) Ullenius, C.; Christenson, B. *Pure Appl. Chem.* **1988**, *60*, 57-64. (d) Christenson, B.; Olsson, T.; Ullenius, C. *Tetrahedron*, **1989**, *45*, 523-534. (e) Nilsson, K.; Ullenius, C.; Krause, N. *J. Am. Chem. Soc.* **1996**, *118*, 4194-4195. (f) Nilsson, K.; Andersson, T.; Ullenius, C.; Gerold, A.; Krause, N. *Chem. Eur. J.* **1998**, *4*, 2051-2058.

[49] Bertz, S. H.; Smith, R. A. *J. Am. Chem. Soc.* **1989**, *111*, 8276-8277.

[50] (a) Krause, N. *J. Org. Chem.* **1992**, *57*, 3509-3512. (b) Krause, N.; Wagner, R.; Gerold, A. *J. Am. Chem. Soc.* **1994**, *116*, 381-382. (c) Canisius, J.; Mobley, T. A.; Berger, S.; Krause, N. *Chem. Eur. J.* **2001**, *7*, 2671-2675.

[51] Sharma, S.; Oehlschlager, A. C. *Tetrahedron* **1989**, *45*, 557-568.

[52] Vellekoop, A. S.; Smith, R. A. J. *J. Am. Chem. Soc.* **1994**, *116*, 2902-2913.

[53] Eriksson, J.; Davidsson, O. *Organometallics* **2001**, *20*, 4763-4765.

[54] Putau, A.; Brand, H.; Koszinowski, K. *J. Am. Chem. Soc.* **2012**, *134*, 613-622.

[55] *Mass Spectrometry Principles and Applications*; De Hoffmann, E.; Stroobant, V..; Wiley: Chichester, 2002.

[56] Cech, N. B.; Enke, C. G.; *Mass Spectrom. Rev.* **2001**, *20*, 362-387.

[57] Hunt, D. F.; Yates, J. R.; Shabanowitz, J.; Winston, S.; Hauer, C. L. *Proc. Natl. Acad. Sci. USA* **1986**, *83*, 6233-6237.

7   References and Notes

[58] For selected reviews on the application of ESI mass spectrometry to the analysis of organometallic species, see: (a) Plattner, D. A. *Int. J. Mass Spectrom.* **2001**, *207*, 125-144. (b) Chen, P. *Angew. Chem.* **2003**, *115*, 2938-2954; *Angew. Chem. Int. Ed.* **2003**, *42*, 2832-2847. (c) Henderson, W.; McIndoe, J. S. *Mass Spectrometry of Inorganic, Coordination and Organometallic Compounds: Tools, Techniques, Tips*, Wiley: Chichester, 2005, pp 175-219. (d) Müller, C. A.; Markert, C.; Teichert, A. M.; Pfaltz, A. *Chem. Comm.* **2009**, 1607-1618.

[59] Koszinowski, K. *J. Am. Chem. Soc.* **2010**, *132*, 6032-6040.

[60] Fleckenstein, J. E.; Koszinowski, K. *Organometallics* **2011**, *30*, 5018-5026.

[61] (a) Yamashita, M.; Fenn, J. B. *J. Phys. Chem.* **1984**, *88*, 4451-4459. (b) Yamashita, M.; Fenn, J. B. *J. Phys. Chem.* **1984**, *88*, 4671-4675.

[62] (a) Iribarne, J. V.; Thomson, B. A. *J. Chem. Phys.* **1976**, *64*, 2287-2294. (b) Thomson, B. A.; Iribarne, J. V. *J. Chem. Phys.* **1979**, *71*, 4451-4463.

[63] (a) Dole, M.; Mack, L. L.; Hines, R. L.; Mobley, R. C.; Ferguson, L. D.; Alice, M. B. *J. Chem. Phys.* **1968**, *49*, 2240-2249. (b) Schmelzeisen-Redeker, G.; Bütfering, L.; Röllgen, F. W. *Int. J. Mass Spectrom. Ion Processes* **1989**, *80*, 139-150. (c) Nehring, H.; Thiebes, S.; Bütfering, L.; Röllgen, F. W. *Int. J. Mass Spectrom. Ion Processes* **1993**, *128*, 123-132.

[64] Kebarle, P. *J. Mass Spectrom.* **2000**, *35*, 804-817.

[65] Kebarle, P.; Verkerk, U. H. *Mass Spectrom. Rev.* **2009**, *28*, 898–917.

[66] O'Hair, R. A. *J. Chem. Commun.* **2006**, 1469-1481.

[67] For a recent instructive example, see: Schröder, D.; Ducháčková, L.; Tarábek, J.; Karwowska, M.; Fijalkowski, K. J.; Ončák, M.; Slavíček, P. *J. Am. Chem. Soc.* **2011**, *133*, 2444-2451.

[68] a) Wang, H.; Agnes, G. R. *Anal. Chem.* **1999**, *71*, 3785-3792. (b) Wang, H.; Agnes, G. R. *Anal. Chem.* **1999**, *71*, 4166-4172. (c) Wortmann, A.; Kistler-Momotova, A.; Zenobi, R.; Heine, M. C.; Wilhelm, O.; Pratsinis, S. E. *J. Am. Soc. Mass Spectrom.* **2007**, *18*, 385-393. (d) Tsierkezos, N. G.; Roithová, J.; Schröder, D.; Ončák, M.; Slavíček, P. *Inorg. Chem.* **2009**, *48*, 6287-6296.

[69] Luedtke, W. D.; Landman, U.; Chiu, Y.-H.; Levandier, D. J.; Dressler, R. A.; Sok, S.; Gordon, M. S. *J. Phys. Chem. A* **2008**, *112*, 9628-9649.

[70] (a) Satterfield, M.; Brodbelt, J. S. *Inorg. Chem.* **2001**, *40*, 5393-5400. (b) Zins, E.-L.; Pepe, C.; Schröder, D. *J. Mass Spectrom.* **2010**, *45*, 1253-1260.

[71] Gabelica, V.; De Pauw, E. *Mass Spectrom. Rev.* **2005**, *24*, 566-587.

[72] Carvajal, C.; Tölle, K. J.; Smid, J.; Szwarc, M. *J. Am. Chem. Soc.* **1965**, *87*, 5548-5553.

[73] *Gaussian 03*, Revision D.01, Frisch, M. J.; Trucks, G. W.; Schlegel, H. B.; Scuseria, G. E.; Robb, M. A.; Cheeseman, J. R.; Montgomery, J. A.; Jr.; Vreven, T.; Kudin, K. N.; Burant, J. C.; Millam, J. M.; Iyengar, S. S.; Tomasi, J.; Barone, V.; Mennucci, B.; Cossi, M.; Scalmani, G.; Rega, N.; Petersson, G. A.; Nakatsuji, H.; Hada, M.; Ehara, M.; Toyota, K.; Fukuda, R.; Hasegawa, J.; Ishida, M.; Nakajima, T.; Honda, Y.; Kitao, O.; Nakai, H.; Klene, M.; Li, X.; Knox, J. E.; Hratchian, H. P.; Cross, J. B.; Bakken, V.; Adamo, C.; Jaramillo, J.; Gomperts, R.; Stratmann, R. E.; Yazyev, O.; Austin, A. J.; Cammi, R.; Pomelli, C.; Ochterski, J. W.; Ayala, P. Y.; Morokuma, K.; Voth, G. A.; Salvador, P.; Dannenberg, J. J.; Zakrzewski, V. G.; Dapprich, S.; Daniels, A. D.; Strain, M. C.; Farkas, O.; Malick, D. K.; Rabuck, A. D.; Raghavachari, K.; Foresman, J. B.; Ortiz, J. V.; Cui, Q.; Baboul, A. G.; Clifford, S.; Cioslowski, J.; Stefanov, B. B.; Liu, G.; Liashenko, A.; Piskorz, P.; Komaromi, I.; Martin, R. L.; Fox, D. J.; Keith, T.; Al-Laham, M. A.; Peng, C. Y.; Nanayakkara, A.; Challacombe, M.; Gill, P. M. W.; Johnson, B.; Chen, W.; Wong, M. W.; Gonzalez, C.; Pople, J. A. Gaussian, Inc., Wallingford CT, **2004**.

[74] Hu, H.; Snyder, J. P. *J. Am. Chem. Soc.* **2007**, *129*, 7210-7211.

[75] Becke, A. D. *J. Chem. Phys.* **1993**, *98*, 5648-5652.

119

7 References and Notes

[76] Leininger, T.; Nicklass, A.; Stoll, H.; Dolg, M.; Schwerdtfeger, P. *J. Chem. Phys.* **1996**, *105*, 1052-1059.

[77] Rassolov, V. A.; Pople, J. A.; Ratner, M. A.; Windus, T. L. *J. Chem. Phys.* **1998**, *109*, 1223-1229.

[78] Figgen, D.; Rauhut, G.; Dolg, M.; Stoll, H. *Chem. Phys.* **2005**, *311*, 227-244.

[79] Møller, C.; Plesset, M. S. *Phys. Rev.* **1934**, *46*, 618-622.

[80] Perdew, J. P.; Burke, K.; Wang, Y. *Phys. Rev. B* **1996**, *54*, 16533-16539.

[81] Adamo, C.; Barone, V. *J. Chem. Phys.* **1998**, *108*, 664-675.

[82] Glendening, E. D.; Reed, A. E.; Carpenter, J. E.; Weinhold, F. *NBO*, Version 3.1, University of Wisconsin, Madison WI, 1993.

[83] Hamann, C. H.; Vielstich, W. *Elektrochemie*, 4th ed., Wiley-VCH, 2005, Weinheim.

[84] Pearson, R. G.; Gregory, C. D. *J. Am. Chem. Soc.* **1976**, *98*, 4098-4104.

[85] For a low-temperature NMR spectroscopic study of mixed LiCu(Me)^{n}Bu·LiCN, see: Lipshutz, B. H.; Kozlowski, J. A.; Wilhelm, R. S. *J. Org. Chem.* **1984**, *49*, 3943-3949.

[86] Davies, R. P.; Hornauer, S.; White, A. J. P. *Chem. Commun.* **2007**, 304-306.

[87] Lide, O. R. *CRC Handbook of Chemistry and Physics*, 88th ed., CRC Press, 2008, Boca Raton.

[88] Watanabe, K.; Yamagiwa, N.; Torisawa, Y. *Organic Process Research & Development* **2007**, *11*, 251-258.

[89] (a) Bhattacharyya, D. N.; Lee, C. L.; Smid, J.; Szwarc, M. *J. Phys. Chem.* **1965**, *69*, 608-611. (b) Böhm, L. L.; Schvlz, G. V. *Ber. Bunsen Ges. Phys. Chem.* **1969**, *73*, 260-264. (c) Chen, Z.; Hojo, M. *J. Phys. Chem. B* **1997**, *101*, 10896-10902.

[90] At higher, and thus, synthetically relevant concentrations the fraction of free ions will be lower because of increased ion pairing according to the law of mass action. This effect can be directly seen in the concentration dependence of the molar conductivity of LiCuPh$_2$·LiCN, see Figure 4.1.3.1

[91] Lipshutz, B. H.; Kozlowski, J. A.; Breneman, C. M. *J. Am. Chem. Soc.* **1985**, *107*, 3197-3204.

[92] Hope, H.; Oram, D.; Power, P. P. *J. Am. Chem. Soc.* **1984**, *106*, 1149-1150.

[93] DePuy, C. H.; Gronert, S.; Barlow, S. E.; Bierbaum, V. M.; Damrauer, R. *J. Am. Chem. Soc.* **1989**, *111*, 1968-1973.

[94] Ingold, C. K. *Structure and Mechanism in Organic Chemistry*, 2nd ed., Cornell University Press: Ithaca, 1969, pp. 428-448.

[95] Kondo, Y.; Matsudaira, T.; Sato, J.; Murata, N.; Sakamoto, T. *Angew. Chem.* **1996**, *108*, 818-820; *Angew. Chem. Int. Ed. Engl.* **1996**, *35*, 736-738.

[96] (a) Langham, W.; Brewster, R. Q.; Gilman, H. *J. Am. Chem. Soc.* **1941**, *63*, 545-549. (b) Bailey, W. F.; Patricia, J. J. *J. Organomet. Chem.* **1988**, *352*, 1-46. (c) Krasovskiy, A.; Straub, B. F.; Knochel, P. *Angew. Chem.* **2006**, *118*, 165-169; *Angew. Chem. Int. Ed.* **2006**, *45*, 159-162.

[97] (a) Moret, M.-E.; Serra, D.; Bach, A.; Chen, P. *Angew. Chem.* **2010**, *122*, 2935-2939; *Angew. Chem. Int. Ed.* **2010**, *49*, 2873-2877. (b) Serra, D.; Moret, M.-E.; Chen, P. *J. Am. Chem. Soc.* **2011**, *133*, 8914-8926.

[98] Fuoss, R. M.; Kraus, C. A. *J. Am. Chem. Soc.* **1933**, *55*, 2387-2399.

[99] Karlström, A. S. E.; Bäckvall, J.-E. *Chem. Eur. J.* **2001**, *7*, 1981-1989.

[100] Norinder, J.; Bäckvall, J.-E.; Yoshikai, N.; Nakamura, E. *Organometallics* **2006**, *25*, 2129-2132.

7 References and Notes

[101] Mori, S.; Nakamura, E.; Morokuma, K. *J. Am. Chem. Soc.* **2000**, *122*, 7294-7307.

[102] Yamanaka, M.; Kato, S.; Nakamura, E. *J. Am. Chem. Soc.* **2004**, *126*, 6287-6293.

[103] Pratt, L. M.; Voit, S.; Mai, B. K.; Nguyen, B. H. *J. Phys. Chem. A* **2010**, *114*, 5005-5015.

[104] Bordwell, F. G.; Bares, J. E.; Bartmess, J. E.; McCollum, G. J.; Van der Puy, M.; Vanier, N. R.; Matthews, W. S. *J. Org. Chem.* **1977**, *42*, 321-325.

[105] Bordwell, F. G.; Van der Puy, M.; Vanier, N. R. *J. Org. Chem.* **1976**, *41*, 1885-1886.

[106] Phillips, W. D.; Rowell, J. C.; Weismann, S. I. *J. Chem. Phys.* **1960**, *33*, 626-627.

[107] Webster, O. W.; Mailer, W.; Benson, R. E. *J. Am. Chem. Soc.* **1962**, *84*, 3678-3684.

[108] Han, Y. K.; Park, J. M.; Choi, S. K. *J. Polym. Sci. Pol. Chem.* **1982**, *20*, 1549-1557.

[109] Ashby, E. C.; Watkins, J. J. *J. Am. Chem. Soc.* **1977**, *99*, 5312-5317

[110] Casitas, A.; King, A. E.; Parella, T.; Costas, M.; Stahl, S. S.; Ribas, X. *Chem. Sci.* **2010**, *1*, 326-330.

[111] (a) Phipps, R. J.; Grimster, N. P.; Gaunt, M. J. *J. Am. Chem. Soc.* **2008**, *130*, 8172-8174. (b) Phipps, R. J.; Gaunt, M. J. *Science*, **2009**, *323*, 1593-1597 (c)Ciana, C.-L.; Phipps, R. J.; Brandt, J. R.; Meyer, F.-M.; Gaunt, M. J. *Angew. Chem. Int. Ed.* **2011**, *50*, 458-462. (d)Duong, H. A.; Gilligan, R. E.; Cooke, M. L.; Phipps, R. J.; Gaunt, M. J. *Angew. Chem. Int. Ed.* **2011**, *50*, 463-466.

[112] (a) Gütz, C.; Lützen, A. *Synthesis* **2010**, 85-90. (b) Dick, G. R.; Woerly, E. M.; Burke, M. D. *Angew. Chem.* **2012**, *124*, 2721 –2726.

[113] Lipton, M. F.; Sorensen, C. M.; Sadler, A. C.; Shapiro, R. H. *J. Organomet. Chem.* **1980**, *186*, 155-158.

[114] Viton, F.; Landreau, C.; Rustidge, D.; Robert, F.; Williamson, G.; Barron, D. *Eur. J. Org. Chem.* **2008**, 6069-6078.

[115] Gajewski, J. J.; Peterson, K. B.; Kagel, J. R.; Huang, Y. C. *J. Am. Chem. Soc.* **1989**, *111*, 9078-9081

[116] Dickinson, C. L., Jr.; Wiley, D. W.; McKusick, B. C. *J. Am. Chem. Soc.* **1960**, *82*, 6132-6136.

[117] Khairallah, G. N.; Thum, C.; O'Hair, R. A. J. *Organometallics* **2009**, *28*, 5002 – 5011.

yes
i want morebooks!

Buy your books fast and straightforward online - at one of world's fastest growing online book stores! Environmentally sound due to Print-on-Demand technologies.

Buy your books online at
www.get-morebooks.com

Kaufen Sie Ihre Bücher schnell und unkompliziert online – auf einer der am schnellsten wachsenden Buchhandelsplattformen weltweit! Dank Print-On-Demand umwelt- und ressourcenschonend produziert.

Bücher schneller online kaufen
www.morebooks.de

VDM Verlagsservicegesellschaft mbH
Heinrich-Böcking-Str. 6-8
D - 66121 Saarbrücken

Telefon: +49 681 3720 174
Telefax: +49 681 3720 1749

info@vdm-vsg.de
www.vdm-vsg.de

Printed by Books on Demand GmbH, Norderstedt / Germany